The Agroecosystem Services of Leguminous Crops

Abd El-Hafeez Zohry • Samiha A. H. Ouda

The Agroecosystem Services of Leguminous Crops

Sustainable Intensification under a Changing Climate

Springer

Abd El-Hafeez Zohry
Field Crops Research Institute, Agricultural Research Center
Crops Intensification Research Department
Giza, Egypt

Samiha A. H. Ouda
Soils, Water and Environment Research Institute, Agricultural Research Center
Water Requirements and Field Irrigation Research Department
Giza, Egypt

ISBN 978-3-032-32460-3 ISBN 978-3-032-32461-0 (eBook)
https://doi.org/10.1007/978-3-032-32461-0

This Springer imprint is published by the registered company Springer Nature Switzerland AG
The registered company address is: Gewerbestrasse 11, 6330 Cham, Switzerland

Preface

Agriculture is currently facing one of the most critical periods in human history. Rapid population growth, declining natural resources, soil degradation, biodiversity loss, water scarcity, and accelerating climate change are placing unprecedented pressure on global food production systems. The challenge confronting modern agriculture is no longer limited to increasing crop productivity; rather, it involves achieving sustainable intensification in ways that maintain ecosystem integrity, enhance resilience, and safeguard environmental resources for future generations.

Within this context, leguminous crops have emerged as essential components of sustainable agricultural systems because of their unique multifunctional contributions to agroecosystem services. Leguminous crops are the gift of nature that might appear to safeguard the environment and ensure win–win benefits for society, economy, and environment. Beyond their traditional role as food and feed crops, legumes contribute significantly to biological nitrogen fixation, nutrient cycling, soil fertility improvement, carbon sequestration, climate regulation, biodiversity enhancement, biological pest control, and water conservation. Their integration into agricultural systems through intercropping, crop rotations, sequence cropping, and cover cropping offers practical and environmentally sound solutions for improving productivity while reducing dependence on external inputs.

The importance of legumes has become even more evident under current climate change scenarios. Increasing temperatures, irregular rainfall patterns, and more frequent drought and heat stress events threaten agricultural sustainability, particularly in arid and semi-arid regions. Legume-based systems have demonstrated considerable potential to improve resilience through enhanced soil health, improved water-use efficiency, reduced greenhouse gas emissions, and greater agroecosystem stability. Consequently, legumes are increasingly recognized as key components of climate-smart agriculture and nature-based solutions for sustainable development.

This book, *The Agroecosystem Services of Leguminous Crops: Sustainable Intensification under a Changing Climate*, was prepared to provide a comprehensive scientific overview of the diverse agroecosystem services provided by legumes and their importance in supporting resilient agricultural systems. The book synthesizes current scientific knowledge regarding the regulating and supporting services of legumes, with particular emphasis on their role in sustainable intensification, ecological resilience, and climate adaptation.

Special attention is given throughout the book to the incorporation of legumes into diversified cropping systems, including intercropping, crop sequencing, crop rotation, and cover cropping. The book also highlights the interactions between legumes and soil biological processes, carbon dynamics, nutrient cycling, water productivity, and biodiversity conservation. Furthermore, considerable emphasis is placed on evidence generated under Egyptian agroecosystem conditions, where increasing pressure on water and soil resources makes sustainable agricultural management particularly important.

The chapters presented in this volume integrate findings from global and regional studies to demonstrate how legumes can contribute to improving productivity while simultaneously supporting ecosystem functioning and environmental sustainability. The book further addresses major challenges facing the adoption of legume-based systems and discusses opportunities for future research, technological innovation, and policy development.

This book is intended for researchers, graduate students, agronomists, soil scientists, ecologists, extension specialists, policymakers, and all readers interested in sustainable agriculture and climate-smart farming systems. It is our hope that this work will contribute to greater understanding of the ecological and agronomic importance of legumes and encourage their broader integration into agricultural systems worldwide.

We sincerely hope that this book will serve as a valuable scientific resource supporting future efforts toward sustainable intensification, food security, environmental conservation, and climate resilience.

Giza, Egypt Abd El-Hafeez Zohry
Giza, Egypt Samiha A. H. Ouda

Competing Interests The authors have no competing interests to declare that are relevant to the content of this manuscript.

Contents

Chapter 1
The Agroecosystem Services of Leguminous Crops: An Overview

Abstract Agroecosystems are human-managed systems that sustain food production but increasingly face challenges related to soil degradation, biodiversity loss, climate variability, and declining resource-use efficiency. Integrating ecological processes into agricultural management is therefore essential for maintaining productivity while safeguarding environmental integrity. Leguminous crops play a central role in enhancing agroecosystem sustainability by delivering a wide range of regulating and supporting services. Through biological nitrogen fixation, legumes reduce dependence on synthetic fertilizers, improve soil fertility, and enhance nutrient cycling. Their high biomass production and extensive root systems contribute to carbon sequestration, improved soil structure, and greater water-holding capacity, thereby mitigating greenhouse gas emissions and strengthening resilience to climate stress. When incorporated through intercropping, crop rotation, or cover cropping, legumes increase biodiversity, stabilize yields, suppress weeds, and promote biological pest control by supporting beneficial arthropod communities. Evidence from Egypt and other regions demonstrates that legume-based cropping systems significantly improve soil organic carbon, nutrient availability, and overall system productivity, particularly under intensive and water-limited conditions. This chapter synthesizes current knowledge on the regulating and supporting services provided by legumes, with special emphasis on Egyptian agroecosystems, and highlights their role as a cornerstone of sustainable intensification and climate-smart agriculture.

Keywords Intercropping · Sequence cropping · Cover crop · Regulating services · Support services

Introduction

Agroecosystems are ecosystems deliberately managed by humans for the production of food, feed, fiber, and other agricultural outputs (Dornbush & von Haden, 2017). In contrast to natural ecosystems, agroecosystems are characterized by

A. E. Zohry, S. A. H. Ouda, *The Agroecosystem Services of Leguminous Crops*,
https://doi.org/10.1007/978-3-032-32461-0_1

simplified species composition and a strong reliance on anthropogenic inputs such as mineral fertilizers, pesticides, mechanization, and irrigation to maintain high levels of productivity (Raj et al., 2021). Although these inputs have contributed significantly to yield enhancement and food security, their intensive and often unsustainable use has accelerated soil degradation, reduced biodiversity, disrupted nutrient cycling, and impaired key ecological processes that sustain long-term agricultural productivity (Hossain et al., 2020; Krasilnikov et al., 2022). Climate change compounds these pressures through rising temperatures, altered precipitation regimes, and increased frequency of extreme weather events, resulting in reduced crop yields, lower nutritional quality, and heightened production risks (Kumari et al., 2025; Wheeler & von Braun, 2013). Consequently, modern agriculture faces the dual challenge of maintaining productivity while minimizing environmental degradation.

The concept of agroecosystem services provides an integrative framework for assessing agricultural systems beyond yield-based metrics. Agroecosystem services encompass the wide range of benefits humans derive from agricultural landscapes, including regulating services such as carbon sequestration, climate regulation, soil conservation, and waste decomposition, as well as supporting services such as soil fertility maintenance, nutrient cycling, biological control, pollination, and weed regulation (Power, 2010). These services are fundamental to ecosystem functioning and form the ecological foundation upon which sustainable agricultural productivity depends. Recent studies emphasize that neglecting these services can lead to yield instability and increased vulnerability of agroecosystems to environmental stressors (Liu et al., 2022; Huang et al., 2015).

Agricultural productivity and the delivery of agroecosystem services are strongly influenced by the management of interconnected resources, including human labor, water, soil, and climate (Banerjee et al., 2020). Sustainable intensification seeks to enhance agricultural outputs while preserving or improving ecosystem services through the strategic management of biological interactions and ecological processes (Jhariya et al., 2021). This can be achieved either by introducing functional biological components, such as diversified cropping systems, or by enhancing ecosystem processes such as nutrient cycling and biological regulation. However, maintaining ecosystem integrity while meeting growing food demands remains one of the most significant challenges facing global agriculture (Kumar et al., 2020). Climate variability further exacerbates this challenge by increasing uncertainty in the management of water and soil resources, particularly in arid and semi-arid regions (Khan et al., 2021).

Agroecosystem services include both positive contributions (services) and negative impacts (disservices) generated by agricultural activities (Liu et al., 2022). Understanding these interactions is essential for developing resilient and sustainable farming systems that support human well-being (Huang et al., 2015). Regulating services provided by agroecosystems include carbon sequestration, climate regulation, waste decomposition, and soil conservation (Liu et al., 2022). Carbon sequestration occurs primarily through crop photosynthesis and the accumulation of soil organic matter, which improves soil structure and mitigates atmospheric CO_2

concentrations (Zabala et al., 2021). Climate regulation is influenced by crop transpiration and evaporation from irrigated surfaces, which can moderate microclimatic temperatures and reduce heat stress (Kumar et al., 2017). Waste decomposition is mediated by soil microbial communities that break down plant residues and organic amendments, thereby recycling nutrients and enhancing soil fertility (Zhang et al., 2007; Bardgett & van der Putten, 2014). Soil conservation is promoted through practices such as cover cropping, which protects soil surfaces, reduces erosion, and improves aggregate stability (Huang et al., 2015).

Supporting services represent another critical dimension of agroecosystem functioning and include soil fertility maintenance, nutrient cycling, biological control, pollination, and weed management (Liu et al., 2022). Soil structure and fertility depend largely on soil aggregation and organic matter content, which regulate nutrient availability and water retention for crop uptake (Zabala et al., 2021). Agroecosystems facilitate the cycling of essential nutrients, particularly nitrogen and phosphorus, through interactions among soil microorganisms, crops, and organic inputs (Rana et al., 2018; Drinkwater & Snapp, 2007). Biological control is provided by natural enemies, including predators and parasitoids, that suppress pest populations and reduce reliance on chemical pesticides (Zhang et al., 2007). Pollinators such as bees and butterflies play a vital role in crop fertilization, directly influencing yield stability and quality (Huang et al., 2015; Klein et al., 2007). Weed regulation is enhanced by seed predators, including birds, ants, beetles, and crickets, which reduce weed seed banks and limit weed establishment in crop fields (Schumacher et al., 2020).

Legumes play a pivotal role in enhancing agroecosystem services by contributing to biological nitrogen fixation, carbon sequestration, improved soil health, and increased resilience to climate variability (Kokkini et al., 2025). The integration of legumes into cropping systems improves soil fertility, reduces dependence on synthetic nitrogen fertilizers, and enhances system sustainability (Lulie, 2017). Through symbiotic associations with rhizobia, legumes convert atmospheric nitrogen into plant-available forms, supplying nitrogen to both the legume and subsequent crops (Stagnari et al., 2017). This process significantly reduces the need for energy-intensive synthetic fertilizers and associated greenhouse gas emissions (Jensen et al., 2010). Legume-based systems have been shown to reduce emissions of CO_2 and N_2O compared with systems reliant on mineral nitrogen fertilization (Hassen et al., 2017). Additionally, legumes improve soil structure, enhance microbial diversity, increase water-holding capacity, and promote carbon sequestration through extensive root systems, thereby reducing fossil energy inputs and increasing agroecosystem resilience (Stagnari et al., 2017).

This chapter synthesizes current knowledge on the regulating and supporting services provided by legumes. It examines the role of legumes in delivering key regulating and supporting services that underpin sustainable agricultural systems. Furthermore, it highlights how the integration of legumes into cropping systems improves soil health, productivity, and climate resilience, with particular emphasis on evidence from legume-based systems in Egypt that demonstrate their contribution to sustainable intensification.

Incorporation of Leguminous Crops into the Cropping System

Diversified cropping systems promote complementary resource use among species, reduce ecological niches for pests and weeds, and enhance system stability under variable environmental conditions (Stagnari et al., 2017). Increasing crop diversity within agricultural systems enhances overall crop productivity while simultaneously improving soil health through interspecific interactions occurring both above and below ground (Rakshit et al., 2018). Aboveground diversification influences canopy architecture, light interception, and microclimatic conditions, which can improve photosynthetic efficiency and regulate canopy temperature (King & Hofmockel, 2017). Diverse canopies also contribute to more efficient radiation use and reduced evaporative demand, thereby enhancing crop performance under water-limited environments (Li et al., 2014).

Belowground diversification plays a critical role in optimizing the utilization of soil water and macro- and micronutrients by fostering diverse root architectures and stimulating soil microbial abundance and activity (King & Hofmockel, 2017). Leguminous species, in particular, contribute to increased microbial biomass and enzymatic activity through root exudation and symbiotic associations with nitrogen-fixing bacteria, which enhance nutrient availability and soil biological functioning (Drinkwater et al., 1998). These interactions improve nutrient cycling efficiency, soil aggregation, and organic matter accumulation, thereby strengthening soil structure and resilience to degradation (Cong et al., 2015).

The incorporation of leguminous crops into cropping systems can be achieved through several agronomic strategies, including intercropping with non-legume species, growing legumes as preceding crops in rotation, or using them as cover crops during fallow periods. Intercropping legumes with cereals enhances nitrogen transfer, improves land-use efficiency, and stabilizes yields under environmental stress (Li et al., 2014). Rotational inclusion of legumes enriches soil nitrogen pools and provides residual benefits to subsequent crops, reducing the need for synthetic nitrogen fertilizers (Peoples et al., 2009). When used as cover crops, legumes protect soil surfaces, suppress weeds, enhance soil organic carbon, and improve water infiltration, contributing to long-term agroecosystem sustainability (Snapp et al., 2019).

Intercropping Leguminous Crops with Non-leguminous Crops

Intercropping, a method of agricultural diversification including the simultaneous cultivation of various crops in the same field, is suggested as an effective way to enhance beneficial arthropods and diminish pest populations (Brooker et al., 2015). Furthermore, intercropping is increasingly recognized as a form of "insurance" against extreme weather due to its ability to enhance system resilience (Akchaya et al., 2025a). Increasing plant diversity by intercropping promotes habitat heterogeneity and resource partitioning, which supports beneficial arthropod communities such as predators and parasitoids while suppressing herbivorous pest populations. Diverse cropping systems exhibit greater functional redundancy and complementarity, allowing them to buffer yield losses under variable weather conditions, and drought stress (Akchaya et al., 2025b). This enhanced resilience arises from differences in rooting depth, growth dynamics, and resource-use efficiency among component crops, which collectively stabilize productivity and reduce risk in agroecosystems (Lin, 2011). This ecological regulation reduces pest outbreaks and dependence on chemical pesticides, contributing to more environmentally sustainable crop protection (Brooker et al., 2015; Letourneau et al., 2011).

Intercropping legumes with other crops can optimize resource utilization and augment overall productivity (Kokkini et al., 2025). Legume intercropping is an agricultural method that utilizes the distinct ecosystem services provided by legumes to improve the resilience and productivity of cropping systems (Landschoot et al., 2024). It enhances biodiversity by supporting beneficial insects, such as pollinators, which can improve crop yield (Rakotomalala et al., 2023). Moreover, nitrogen fixation by legumes diminishes the need for synthetic fertilizers, enhances nutrient cycling efficiency, and bolsters soil health (Hufnagel et al., 2020), as well as aids in weed suppression (Gazoulis et al., 2025b).

A notable example of legume intercropping is its integration with cereals. Intercropping legumes with cereals has several advantages for resource management and production improvement (Gaba et al., 2015). This method may result in enhanced resource use, reduced soil erosion, and augmented soil fertility (Yu et al., 2016). Furthermore, it increases nitrogen and carbon content in the rhizospheric soil, and therefore they can be further used by successive crops (Zang et al., 2015). Xiong et al. (2013) reported increased iron content in the rhizosphere of a peanut intercropping system with maize, in which iron enrichment enhanced the carbon and nitrogen metabolism as well as photosynthetic efficiency of the peanut crop and the resistance of both crops against various environmental stresses.

Intercropping can augment yields on limited available lands, as observed in Egypt, where the intercropping soybean with maize (El-Shamy et al., 2015) and intercropping faba bean with wheat (Zohry & Ouda, 2025) facilitates the production of soybean and faba bean without necessitating more space for cultivation of these two crops. Moreover, intercropping provides risk mitigation for farmers, as one intercrop can offset the potential failure of another within the system (Martin-Guay et al., 2018). Intercropping legumes with non-legumes may enhance income per labor unit (Li et al., 2020).

Leguminous Crops Preceding Non-leguminous Crops

Sequence cropping emphasizes the arrangement of crops during a brief period, often one year, with diminished reliance on a consistent, recurring pattern (FAO, 2003). Diversified cropping systems, particularly those using legumes, have been suggested to improve food production while minimizing inputs and environmental effects (Rakotomalala et al., 2023). Planting legume crops prior to non-legume crops is a sustainable agricultural method that has various advantages, including enhanced soil health by increasing soil organic matter, enhancing microbial activity, and enriching soil nitrogen pools through biological nitrogen fixation (Weiner et al., 2024). These residual effects improve nutrient availability and soil structure for subsequent crops, leading to increased yields and reduced requirements for synthetic nitrogen fertilizers (Peoples et al., 2009).

The cultivation of legume crops prior to non-legume crops is an advantageous agricultural practice that enhances total crop productivity (Sainio et al., 2019). Numerous prior studies have demonstrated that the production and quality of subsequent harvests are affected by the nature of the preceding crop (Sainio et al., 2019). Faba beans, as preceding crops, can markedly enhance the output and quality of tomatoes, especially regarding calcium content (Raffa et al., 2022). This approach also disrupts insect and disease cycles and improves soil health (Lazali & Drevon, 2023) and diminishes pest and disease pressures, resulting in increased yields for preceding crops, especially non-legumes (Nanganoa et al., 2019).

Utilizing Legumes as Cover Crops

Cover crops are characterized as non-cash crops cultivated prior to or concurrently with primary crops to maintain soil coverage with vegetation for extended periods (Blanco-Canqui et al., 2015). Legumes serve as advantageous cover crops, providing benefits such as nitrogen fixation, weed reduction, and soil enhancement (Kocira et al., 2020). Utilizing legumes as cover crops during fallow times enhances soil health, mitigates erosion and nitrate leaching, promotes water infiltration, and preserves soil moisture (Gazoulis et al., 2025a). As cover crops, legumes protect the soil surface from erosion, suppress weed emergence through competition and shading, and enhance soil fertility via nitrogen fixation and organic matter inputs (Kocira et al., 2020). The decomposition of legume residues releases biologically fixed nitrogen and stimulates soil microbial processes, improving nutrient cycling and soil aggregation (Snapp et al., 2019). Cereal crops typically exhibit increased yields when grown following legume crops, attributed to the nitrogen advantages derived from legume crop residue (Preissel et al., 2015). Guinet et al. (2020) demonstrated that the yield of unfertilized wheat was elevated by 27% and 25% following faba bean and lentil cultivation, respectively, in comparison to the yield achieved after barley. Moreover, wheat production increased by 66% and 51% following summer legumes, in comparison to the yield gained after sorghum.

Collectively, these benefits highlight the critical role of legume-based intercropping, rotations, and cover cropping in promoting resilient, productive, and environmentally sustainable agroecosystems.

Examples of Legume-Based Cropping Systems Implemented in Egypt

Legume–Cereal Intercropping

Intercropping cereals with legumes represents a practical diversification strategy that enhances agroecosystem services in Egyptian farming systems, particularly in the irrigated Nile Valley and newly reclaimed soils. Field experiments conducted in Egypt have shown that dry bean–maize and cowpea–maize intercrops can increase the land equivalent ratio (LER), improve economic returns, and enable partial rationalization of nitrogen fertilizer inputs without causing substantial yield penalties in maize (El-Mehy et al., 2023; El-Ghobashy et al., 2018). Similarly, Zohry and Ouda (2025) reported an increase in wheat yield, as well as land and water equivalent ratios, when intercropped with legume crops. Whereas, Ouda and Zohry (2026) reported similar results when soybean was intercropped with maize.

These locally generated findings are consistent with broader ecological mechanisms. Legumes contribute biologically fixed nitrogen that benefits the cereal component and subsequent crops, while increased plant diversity enhances habitats for natural enemies and reduces pest outbreaks (Peoples et al., 2009; Letourneau et al., 2011). Under Egyptian conditions, temporal niche management practices—such as relay intercropping, changing planting dates, and optimized planting geometry—have proven particularly effective. For example, Egyptian field trials identified specific cowpea–maize spatial arrangements and cowpea sowing dates that maximized both LER and net returns under local irrigation and management regimes (El-Ghobashy et al., 2018; El-Mehy et al., 2023).

Leguminous Crops Preceding Non-leguminous Crops

In Egyptian agriculture, the inclusion of leguminous crops in rotation with non-leguminous crops, particularly cereals, plays a critical role in maintaining soil fertility, enhancing nitrogen availability, and improving crop productivity through increased soil organic matter. Sheha et al. (2016) reported a 9% increase in maize yield when maize was preceded by faba bean compared with wheat or sugar beet. They also observed higher soil nitrogen, phosphorus, and potassium levels following maize harvest when faba bean was the preceding crop. Similarly, Zen El-Dein et al. (2022) found that maize yield increased by 17% when preceded by Egyptian clover compared with wheat. Comparable yield benefits were reported by Gomaa et al. (2021), who observed improved wheat performance and yield when soybean preceded the crop.

Legumes Used as Cover Crops

Egyptian researchers have increasingly focused on the use of legumes as cover crops to improve soil health, enhance crop productivity, and optimize water use under arid conditions. One promising approach is improved fallow, defined as the cultivation of a short season legume between summer and winter crops. In Egypt, the introduction of fahl clover (a variety of Egyptian clover) between maize and wheat has been shown to improve soil fertility and increase the following wheat yield (Zohry et al., 2017). Zohry et al. (2017) further demonstrated that incorporating fahl clover between maize and wheat increased wheat production by 32% and improved system water productivity by 24%. Awad et al. (2023) reported higher soil nitrogen, phosphorus, and potassium contents when wheat was preceded by fahl clover rather than maize, along with an 8% increase in wheat yield compared with maize–wheat sequences. In addition, Ouda et al. (2020) documented a 50% increase in soil nitrogen concentration when fahl clover preceded wheat compared with maize as the preceding crop.

Collectively, these Egypt-based studies indicate that legume–cereal intercropping, legume-based rotations, and legume cover cropping are effective and scalable strategies for increasing productivity while improving soil health and system resilience under water-limited and heat-stressed Egyptian agroecosystems.

Example of a Legume-Based Crop Rotation Implemented in Egypt

The integration of legume-based cropping systems—whether through intercropping, rotation with non-leguminous crops, or the use of leguminous cover crops—plays a crucial role in increasing crop diversification, improving resource-use efficiency, and sustainably enhancing food production in Egyptian agroecosystems. Such systems contribute not only to higher productivity but also to long-term soil fertility and overall agroecosystem functioning.

A comprehensive example of this approach was reported by Abou-Keriasha et al. (2012a), who implemented a three-year legume-based crop rotation system in Middle Egypt. The rotation was divided into three distinct sections, each designed to maximize land use, biological nitrogen inputs, and system productivity through the strategic inclusion of legumes.

In the first section of the rotation, cotton was relay intercropped with wheat in March. Following wheat harvest in April, cowpea was introduced and intercropped with cotton, and both crops were harvested together in September. This sequence allowed continuous ground cover, efficient use of the growing season, and the incorporation of a nitrogen-fixing legume alongside major cash crops (Fig. 1.1a, b).

Fig. 1.1 Cotton relay intercropped with wheat system (**a**) and cowpea intercropped with cotton (**b**)

Fig. 1.2 Faba bean intercropped with wheat (**a**), followed by soybean intercropped with maize (**b**) followed by short season clover (**c**)

The second section of the rotation focused on intensive legume integration across both winter and summer seasons. During winter, faba bean was intercropped with wheat (Fig. 1.2a). In the subsequent summer season, soybean was intercropped with maize at 50% of its recommended planting density, ensuring reduced competition while maintaining the benefits of biological nitrogen fixation (Fig. 1.2b). In addition, short season Egyptian clover was cultivated as an early winter crop following the harvest of soybean–maize intercropping system, functioning as a short season leguminous cover crop that further enhanced soil fertility (Fig. 1.2c).

In the third section of the rotation, faba bean was intercropped with onion during the winter season (Fig. 1.3a), while cowpea was intercropped with maize in the summer season (Fig. 1.3b). This combination diversified crop functional groups and market outputs while maintaining a consistent presence of legumes throughout the rotation cycle.

Fig. 1.3 Onion intercropped with wheat (**a**) and cowpea intercropped with maize (**b**)

Year 1	Year 2	Year 3
Cotton relay intercropped on wheat then cowpea intercropped with cotton	Faba bean intercropped with onion Cowpea intercropped with maize	Faba bean intercropped with wheat Soybean intercropped with maize Fahl clover
Faba bean intercropped with wheat Soybean intercropped with maize Fahl clover	Cotton relay intercropped on wheat then cowpea intercropped with cotton	Faba bean intercropped with onion Cowpea intercropped with maize
Faba bean intercropped with onion Cowpea intercropped with maize	Faba bean intercropped with wheat Soybean intercropped with maize Fahl clover	Cotton relay intercropped on wheat then cowpea intercropped with cotton

Fig. 1.4 Legume-based intensive crop rotation (adapted from Abou-Keriasha et al., 2012a)

The cumulative effects of this legume-intensive rotation were substantial. After completion of the crop rotation cycle, soil organic matter content increased by 38%. Moreover, soil available nitrogen, phosphorus, and potassium levels rose by 43%, 45%, and 35%, respectively. These improvements were largely attributed to the continuous inclusion of leguminous species—namely cowpea, faba bean, soybean, and short season clover—across all sections of the rotation, which enhanced nutrient cycling, organic matter inputs, and overall soil quality. Figure 1.4 illustrates the layout of this intensive crop rotation.

Overall, this Egyptian case study clearly demonstrates the potential of well-designed legume-based crop rotations to improve soil fertility, sustain high productivity, and enhance agroecosystem resilience under intensive and resource-constrained farming conditions.

The Services Provided by Legumes when Incorporated in the Cropping System

Legumes can augment some of the services offered by the agroecosystem, thereby enhancing the resilience of the cropping system. Regarding the regulating services of the agroecosystem, legumes can play a role in carbon sequestration, climate regulation (climate variability and change), and soil conservation, as well as facilitating waste decomposition. Legumes can also contribute to supporting services of the agroecosystem by increasing soil fertility, enhancing nutrient cycling, facilitating biological control.

Regulating Services of Legumes

Carbon Sequestration

Improved agricultural management practices enhance resource-use efficiency and contribute substantially to climate change mitigation, with nearly 90% of the mitigation potential attributed to soil carbon sequestration and only about 10% to direct reductions in greenhouse gas emissions (Ortiz-Monasterio et al., 2010). Within this context, legume-based cropping systems represent one of the most effective nature-based solutions for increasing soil organic carbon stocks in agroecosystems and mitigating climate change impacts (Meena et al., 2015).

Legumes contribute to carbon sequestration primarily through their biological nitrogen fixation capacity, which stimulates plant growth and increases both above- and belowground biomass inputs to the soil (Srivastava et al., 2025a, b). Their extensive and deep root systems enhance rhizodeposition and promote the stabilization of organic carbon in soil aggregates, thereby functioning as an efficient carbon sink (Hossain et al., 2023). In comparison with non-leguminous crops, legumes generally produce higher total biomass, resulting in greater carbon inputs to soil through roots, residues, and exudates (Kumar et al., 2018; Barbieri et al., 2023).

The incorporation of legumes into cropping systems leads to substantial residue return to the soil, providing energy sources for soil microorganisms and enhancing microbial activity and carbon cycling (Yadav et al., 2017). These processes facilitate the transformation of plant-derived carbon into stable soil organic matter fractions. Long-term studies have demonstrated that legume-based intercropping systems consistently enhance soil carbon stocks by increasing rhizodeposition and residual biomass, while simultaneously improving soil fertility through greater nitrogen availability (Yang et al., 2023). Such systems can also modify soil microbial community composition and functionality, promoting greater carbon-use efficiency and long-term carbon stabilization (Wang et al., 2021).

In grassland and mixed farming systems, the inclusion of legumes has been shown to significantly increase soil organic content concentrations compared with non-legume systems, underscoring their role in long-term carbon sequestration (Lüscher et al., 2014). Evidence suggests that legume-inclusive cropping systems can sequester up to 30% more soil organic carbon than systems dominated by non-leguminous species (Kumar et al., 2018). Since increases in soil organic matter are strongly associated with enhanced carbon sequestration, the strategic integration of legumes into agricultural rotations and intercrops represents a viable pathway for improving soil quality and strengthening climate resilience in agroecosystems (Nazir et al., 2024; Lal, 2020; Paustian et al., 2019).

Carbon Sequestration in Egyptian Agroecosystems

In Egyptian agroecosystems, particularly within the intensively cultivated Nile Valley and Delta, improved agricultural management practices are essential for enhancing resource-use efficiency and mitigating climate change impacts. Similar to global trends, the majority of mitigation potential in Egyptian agriculture is associated with soil carbon sequestration rather than direct reductions in greenhouse gas emissions, with soil organic carbon accumulation accounting for a substantial proportion of climate mitigation benefits (Ortiz-Monasterio et al., 2010). However, Egyptian soils are often characterized by low organic matter content due to high temperatures, intensive tillage, residue removal, and continuous cereal-based cropping, making carbon sequestration a critical challenge (FAO, 2019).

Under these conditions, legume-based cropping systems represent a highly effective strategy for enhancing soil organic content stocks and improving soil quality in Egypt. Grain and forage legumes such as faba bean (*Vicia faba* L.), berseem clover (*Trifolium alexandrinum* L.), chickpea (*Cicer arietinum* L.), cowpea (*Vigna unguiculata* L.), and soybean (*Glycine max* L.) are widely adapted to Egyptian irrigated agriculture and play a key role in carbon sequestration through their biological nitrogen fixation capacity and high biomass production (Meena et al., 2015; Abou-Keriasha et al., 2012b). Biological nitrogen fixation enhances plant growth and increases both above- and belowground carbon inputs, thereby stimulating the buildup of soil organic matter (Srivastava et al., 2025a, b; Hossain et al., 2023).

Legumes grown under Egyptian conditions typically develop extensive root systems that contribute significantly to rhizodeposition and root-derived carbon inputs, which are more stable in soil than shoot-derived carbon. These processes promote carbon stabilization in soil aggregates, particularly in alluvial soils of the Nile Delta (Abbady et al., 2016). Studies conducted in Middle and Lower Egypt have shown that legume–cereal rotations and intercropping systems increase total biomass production and residue return to the soil, leading to higher soil organic content levels compared with continuous cereal monocropping (Abou-Keriasha et al., 2012a; Abbady et al., 2016).

In irrigated Egyptian systems, crop residues from legumes provide a critical energy source for soil microorganisms, enhancing microbial activity, nutrient cycling, and carbon transformation processes (Abbady et al., 2016). Long-term legume-based intercropping and rotation systems have been shown to improve soil fertility by increasing soil nitrogen and carbon pools through enhanced rhizodeposition and residual biomass inputs (Abou-Keriasha et al., 2012a, b). These changes can also alter soil microbial community structure and function, increase microbial carbon-use efficiency, and promote long-term carbon stabilization, even under high-temperature conditions typical of arid and semi-arid regions (Wang et al., 2021).

Overall, legume-inclusive cropping systems have the potential to sequester up to 30% more soil organic carbon than systems dominated by non-leguminous crops, contributing to improved soil structure, water-holding capacity, and resilience to climate stress (Kumar et al., 2018; Nazir et al., 2024). Therefore, the strategic integration of legumes into Egyptian cropping systems represents a practical and climate-smart approach for enhancing carbon sequestration, sustaining soil productivity, and supporting long-term agricultural resilience under increasing climate variability.

Decomposition of Crop Residues in the Soil

Crop residue decomposition is a fundamental soil biological process that regulates nutrient cycling, soil organic matter dynamics, and overall soil fertility (Sarkar et al., 2020). Residues derived from leguminous crops generally decompose more rapidly in soil than those of non-leguminous crops, resulting in a faster release of nutrients into the soil solution (Da Silva et al., 2021). This accelerated decomposition is largely attributed to the higher nitrogen concentration and lower lignin-to-nitrogen ratio of legume residues, which make them more chemically labile and more easily decomposed by soil microorganisms (Sarkar et al., 2020).

Once incorporated into the soil, legume residues undergo rapid microbial breakdown, releasing nitrogen, phosphorus, and other essential nutrients that can be readily taken up by subsequent crops. This process improves nutrient synchronization between residue mineralization and crop demand, thereby reducing nutrient limitations and mitigating yield losses caused by abiotic stresses such as drought or temperature extremes (Srivastava et al., 2025a, b). In addition to nutrient supply, the decomposition of legume residues contributes to the formation of soil organic matter, which enhances soil structure, aggregation, and microbial activity (Lucia et al., 2025).

The benefits of legume residue decomposition extend beyond nutrient cycling to include improvements in soil physical properties. Decomposing legume residues increase soil porosity and aggregation, leading to enhanced water-holding capacity, improved soil permeability, and better aeration (Smitha et al., 2019). These changes support deeper root growth and increase crop productivity by improving nutrient availability within the root zone. Moreover, improved soil organic matter content enhances soil resilience by increasing its capacity to retain water and buffer crops against extreme weather events, including droughts and irregular rainfall patterns (Vadez et al., 2012).

Decomposition of Crop Residues in Egyptian Agroecosystems

In Egypt's irrigated Nile Valley and Delta systems, legume residues (from crops such as Egyptian clover, faba bean, cowpea and soybean) are an important local source of organic inputs that can partly remedy the region's generally low soil organic matter. Field studies and reviews from Egyptian agricultural research highlight that integrating legumes into rotations and intercropping systems increases residual soil fertility, organic matter, and productivity under local management conditions (Zohry & Ouda, 2025, Ouda & Zohry, 2023, Ouda et al., 2020, Zohry et al., 2020). For instance, intensive rotations that include legumes have been shown to raise residual soil organic matter and available nitrogen compared with continuous cereal cropping, improving yield stability in irrigated systems (Abou-Keriasha et al., 2012a).

Recent Egyptian studies on berseem and faba bean indicate that these crops not only supply nitrogen via biological fixation but also return substantial amounts of labile residues that decompose under irrigated conditions to quickly benefit following crops (Ouda et al., 2020; Ouda & Zohry, 2023); this is particularly valuable where residue removal is otherwise common. FAO technical guidance on Egyptian clover documents measurable improvements in soil humus and nitrogen following berseem cultivation, reinforcing its role as a soil-building crop in Egyptian farming systems (FAO, 2013).

At the landscape scale, assessments of the Nile Delta show opportunities for soil organic content recovery where legumes and improved management are adopted; long-term cultivation histories in Northern Delta soils point to measurable soil organic content accumulation where crop sequences and cover cropping improve organic inputs, supporting the potential for legumes to contribute to regional soil carbon gains (Arshad et al., 2020).

Soil Conservation

Soil conservation is a key regulating service provided by legumes, particularly through their influence on soil structure, aggregation, and surface protection. Leguminous crops typically possess deep and extensive root systems that physically bind soil particles together, improving aggregate stability and creating a more resilient soil structure (Gulia et al., 2025). This enhanced aggregation reduces soil susceptibility to erosion and compaction while improving porosity and root penetration.

Intercropping systems that include legumes have proven especially effective in controlling soil erosion, particularly when integrated with conservation practices such as mulching, reduced tillage, and cover cropping, which reduce the kinetic energy of rainfall, as well as lower soil erosion losses by up to 50% (Lal, 2017). According to Liu et al. (2025), legume-based intercropping reduces surface runoff by creating a denser and more diverse canopy that intercepts rainfall and enhances soil infiltration capacity, thereby limiting overland water flow.

By minimizing soil erosion, legumes indirectly enhance nutrient availability by reducing nutrient losses associated with sediment transport, which leads to improved soil fertility and long-term soil health (Yuvaraj et al., 2020a). Furthermore, the enhancement of soil aggregation and organic matter content improves water infiltration and water-holding capacity, increasing soil resilience to both drought stress and periods of intense rainfall (Rahman et al., 2022). These properties are particularly valuable under climate variability, where maintaining soil moisture and reducing erosion are essential for sustaining crop productivity.

Soil Conservation in Egyptian Agroecosystems

Soil conservation is particularly relevant in Egypt where fine-textured alluvial soils, irrigation practices, and salinity issues amplify the consequences of poor structure and surface runoff. Egyptian field trials and reviews indicate that inclusion of legumes (notably Egyptian clover and faba bean) in rotations and as cover crops improves surface cover, reduces runoff, and enhances soil structural properties in irrigated plots—contributing to reduced erosion and better retention of productive topsoil (Abou-Keriasha et al., 2012b, 2013).

National assessments and policy documents also stress the importance of soil-building crops for combating land degradation across the Nile Delta; Egypt's recent national action planning and soil quality studies recognize legumes' role in restoring organic matter and improving infiltration as part of integrated land-management strategies to reduce vulnerability to salinization, waterlogging and extreme events (Ministry of Agriculture and Land Reclamation, 2024).

Climate Regulation

Legumes Alleviate Climate Fluctuations

Legume-based cropping systems, particularly when used in intercropping or diversified rotations, contribute to climate regulation by moderating the microclimate within agricultural fields. Intercropping with legumes modifies key microclimatic factors such as light intensity, air and soil temperature, relative humidity, and wind speed, creating more favorable growing conditions for associated crops (Gebru, 2015; Akchaya et al., 2025a). These modifications reduce thermal and moisture stress, which are increasingly important under conditions of climate variability.

The structural diversity introduced by legumes alters light interception and distribution within the canopy. Taller crops can partially shade shorter or shade-tolerant legume species, while legumes contribute to more uniform ground cover that reduces excessive radiation reaching the soil surface (Nyawade et al., 2019). This complementary light use improves canopy photosynthesis while preventing soil overheating and moisture loss. Increased leaf area index in legume-based intercrops also leads to lower canopy and soil temperatures by enhancing shading and reducing evaporative demand (Shumet et al., 2022).

Legumes further influence microclimate regulation by modifying canopy humidity. Through transpiration, legume leaves release water vapor into the canopy, increasing relative humidity and reducing vapor pressure deficit, which can alleviate crop water stress during hot and dry periods (Chia & Lim, 2022). In addition, dense legume cover reduces wind speed near the soil surface, limiting desiccation and erosion (Schnarr et al., 2022). These microclimatic effects collectively enhance soil moisture retention and stabilize crop growth under fluctuating temperature and rainfall regimes.

Beyond immediate microclimatic effects, legumes enhance climate resilience by improving soil structure, organic matter content, and water-holding capacity, thereby buffering crops against short-term climate extremes such as heat waves or irregular rainfall (Rahman et al., 2022). Many legume species possess inherent tolerance to drought and temperature fluctuations due to deep rooting systems and physiological plasticity, making them particularly suitable for environments experiencing increased climatic variability (Dutta et al., 2022).

Relevance to Egyptian Agroecosystems

In Egypt, where cropping systems are exposed to high temperatures, intense solar radiation, and increasing variability in water availability, the microclimatic regulation provided by legumes is especially valuable. Intercropping systems involving Egyptian clover, cowpea, soybean, or faba bean with cereals such as maize or wheat can reduce canopy and soil temperatures under irrigated conditions in the Nile Valley and Delta (Abdel-Wahab & El Manzlawy, 2016; Abdel-Wahab & Abd El-Rahman, 2016; Ouda & Zohry, 2023). Enhanced ground cover limits soil moisture losses from evaporation, while increased canopy humidity mitigates heat stress during critical growth stages (Ouda et al., 2020). These effects are particularly important under climate change scenarios predicting higher temperatures and more frequent heat waves across Egypt.

Legumes Mitigate Climate Change

Legumes play a central role in climate change mitigation by reducing greenhouse gas emissions, enhancing soil carbon storage, and increasing the overall resilience of agricultural systems (Srivastava et al., 2025a, b). One of the most significant mitigation pathways associated with legumes is their ability to biologically fix atmospheric nitrogen through symbiotic associations with rhizobia. This process reduces reliance on synthetic nitrogen fertilizers, whose production and application are energy-intensive and major contributors to greenhouse gas emissions, particularly nitrous oxide (N_2O) (Marteau-Bazouni et al., 2024a; Smitha et al., 2019).

By improving soil fertility through biological nitrogen fixation and residue decomposition, legumes enhance nutrient availability for subsequent crops, reducing the need for external inputs while sustaining productivity under climate stress (Masson-Boivin & Sachs, 2018). The decomposition of legume residues after harvest releases nitrogen and other nutrients in synchrony with crop demand, improving nitrogen-use efficiency and lowering emission losses (Jahanzad et al., 2016). Moreover, diversified cropping systems that include legumes tend to exhibit lower N_2O emissions per unit of yield compared with cereal monocultures, due to improved nutrient cycling and soil structure (Jensen et al., 2012).

At the system level, the integration of legumes into intercropping and rotation strategies contributes to climate-smart agriculture by increasing yield stability, enhancing food security, and reducing environmental risks associated with climate change. These systems diversify production, spread risk, and improve agroecosystem functioning, making them more robust under future climate scenarios (Lüscher et al., 2014; Lal, 2020).

Climate Change Mitigation in Egyptian Agroecosystems

In Egypt, mitigating climate change while maintaining food security is a major agricultural challenge due to limited arable land, intensive irrigation, and rising temperatures. The inclusion of legumes in Egyptian cropping systems offers a practical mitigation pathway by reducing dependence on mineral nitrogen fertilizers, which are widely used in cereal-based systems. Crops such as Egyptian clover and faba bean have long been recognized for their role in improving soil fertility and reducing fertilizer requirements in subsequent cereal crops (Ouda et al., 2020; Ouda & Zohry, 2023; Zohry & Ouda, 2025). By lowering input-related emissions and enhancing soil carbon and nitrogen cycling, legume-based systems can contribute meaningfully to Egypt's national climate mitigation and adaptation strategies while sustaining productivity in the Nile Valley and Delta.

Support Services of Legumes

Soil Fertility and Nutrient Cycling

Legumes play a fundamental role in supporting soil fertility and nutrient cycling due to their unique capacity to biologically fix atmospheric nitrogen, thereby enriching soil nitrogen pools and reducing dependence on synthetic nitrogen fertilizers (Srivastava et al., 2025a, b). Legumes achieve nitrogen fixation through symbiotic associations with rhizobia bacteria housed in root nodules, where atmospheric nitrogen is converted into plant-available forms, supporting both crop growth and soil nutrient pools (Stomph et al., 2020b). This biological nitrogen fixation not only supplies nitrogen directly to the legume crop but also contributes residual nitrogen to subsequent or associated non-legume crops, enhancing overall system fertility and nutrient-use efficiency (Marteau-Bazouni et al., 2024b). Beyond nitrogen inputs, legumes improve soil fertility through their positive effects on soil structure, organic matter accumulation, and nutrient availability (Akchaya et al., 2025b). These benefits arise from the production of extensive root systems and nitrogen-rich residues that decompose readily, increasing soil organic carbon and improving soil aggregation (Stomph et al., 2020a).

In addition, legume cultivation promotes a more diverse and active soil microbial community, which is critical for nutrient mineralization, nitrogen cycling, and phosphorus mobilization (Marteau-Bazouni et al., 2024b). A diversified and dynamic soil microbiome enhances soil biological activity and ecosystem functioning, thereby improving nutrient availability and strengthening the resilience and productivity of cultivated crops under variable environmental conditions (Yuvaraj et al., 2020b).

Enhancing Soil Fertility in Egyptian Agroecosystems

Under Egyptian agroecosystem conditions, where soils are often characterized by low organic matter content and declining fertility due to intensive cultivation and high mineral fertilizer inputs, legumes play a particularly critical role in enhancing soil fertility and nutrient cycling. Several studies conducted under Nile Valley and newly reclaimed desert soils have demonstrated that the inclusion of grain and forage legumes in cropping systems significantly increases soil available nitrogen and improves nutrient balance for subsequent cereal crops (Zohry et al., 2020; Abou-Keriasha et al., 2012a, b). In legume–cereal intercropping and rotation systems, biologically fixed nitrogen contributes not only to the immediate legume crop but also to companion or following crops through root exudates, residue decomposition, and enhanced microbial turnover, thereby reducing the need for synthetic nitrogen fertilizers (Ouda et al., 2020). Moreover, legumes grown under Egyptian conditions have been shown to enhance phosphorus availability through rhizosphere acidification and stimulation of phosphate-solubilizing microorganisms, improving the efficiency of native and applied phosphorus fertilizers in calcareous soils (Ouda & Zohry, 2026; Zohry & Ouda, 2025; El-Dessougi et al., 2018). The incorporation of legume residues further promotes soil organic carbon accumulation and microbial activity, accelerating nutrient mineralization processes and improving overall soil nutrient cycling. These effects are particularly important in water-limited environments, where improved nutrient availability and soil biological activity enhance crop resilience and sustain productivity under climatic and resource constraints typical of Egyptian agriculture.

Legumes Contribute to Biological Control

Legumes play a pivotal role in biological control within agroecosystems by enhancing the abundance, diversity, and activity of natural enemies of crop pests, thereby contributing to ecologically based pest management (Pierre et al., 2021). The incorporation of legumes into cropping systems—particularly through intercropping, crop rotation, or cover cropping—creates more structurally and functionally diverse habitats that support predators and parasitoids of major agricultural pests (Pierre et al., 2023). Legume species often provide floral resources such as nectar and pollen, alternative prey, and shelter, which are essential for sustaining populations of beneficial arthropods throughout the cropping season (Gontijo, 2019). Consequently, legume-based systems frequently exhibit higher densities of key natural enemies, including lady beetles, lacewings, spiders, predatory bugs, and hymenopteran parasitoids, leading to enhanced suppression of herbivorous insect populations (Pierre et al., 2021).

In addition, legumes used as cover crops or intercrops can indirectly reduce pest pressure by improving soil health and crop vigor, thereby increasing plant tolerance to pest damage. Improved soil structure, enhanced microbial activity, and increased nitrogen availability associated with legumes can result in more resilient crops that are less susceptible to pest outbreaks (Rodríguez et al., 2022). Legume cover crops also suppress weeds that may otherwise serve as alternate hosts for insect pests, further disrupting pest life cycles and reducing infestation levels (Dingha et al., 2021). Collectively, these mechanisms highlight the multifunctional role of legumes in promoting biological control while reducing dependence on synthetic pesticides and supporting sustainable pest management strategies.

Relevance to Egyptian Agroecosystems

In Egyptian agroecosystems, where intensive monocropping and high pesticide inputs are common—particularly in cereal- and vegetable-based production systems—the integration of legumes offers a practical and environmentally sound approach to enhancing biological control. Legume–cereal intercropping systems, such as cowpea-maize (Hamd-Alla et al., 2014) and faba bean-tomato (Elsadany & El-Shamy, 2014) have been shown to increase populations of beneficial insects and reduce infestations of key pests, including aphids, stem borers, and leaf-feeding insects under Nile Delta and Middle Egypt conditions. Egyptian studies have demonstrated that the inclusion of legumes in cropping sequences improves agroecosystem diversity and stability, which is critical for sustaining natural enemy communities in hot, arid environments characterized by strong seasonal fluctuations (Metwally et al., 2017).

Moreover, the widespread cultivation of leguminous forage crops such as Egyptian clover provides continuous floral and habitat resources that support parasitoids and predators during and between cropping seasons, thereby strengthening natural pest regulation at the landscape level (Bondok & Saad-Allah, 2024). This is particularly relevant for smallholder farming systems in Egypt, where reducing pesticide costs and minimizing environmental contamination are national priorities. By integrating legumes into existing cropping systems, Egyptian agriculture can enhance biological control services, improve system resilience to pest outbreaks, and align with sustainable intensification goals under increasing climatic and ecological pressures.

Conclusion

The integration of legumes into cropping systems offers a powerful and multifunctional approach to enhancing agroecosystem services while sustaining agricultural productivity under increasing environmental and climatic pressures. Across a wide range of management strategies—including intercropping, crop sequencing, and the use of leguminous cover crops—legumes consistently improve soil fertility, nutrient cycling, carbon sequestration, and biological regulation. Their ability to biologically fix atmospheric nitrogen reduces reliance on synthetic fertilizers, lowers greenhouse gas emissions, and strengthens soil health, forming a critical foundation for long-term agricultural sustainability.

Evidence synthesized in this chapter demonstrates that legume-based systems enhance biodiversity, support natural enemies of crop pests, suppress weeds, and improve yield stability, particularly in intensive production systems. Findings from Egyptian agroecosystems clearly show that legumes play a vital role in restoring soil organic matter, improving nutrient availability, increasing water productivity, and buffering crops against heat and water stress under arid and semi-arid conditions. These benefits are especially important in Egypt, where limited arable land, intensive irrigation, and climate change pose growing challenges to food security.

Overall, legume-based cropping systems represent a practical, scalable, and nature-based solution for sustainable intensification. Their broader adoption can reconcile productivity goals with environmental stewardship, contributing to resilient agroecosystems that support food security, climate adaptation, and long-term soil conservation. Strategic integration of legumes should therefore be considered a central component of future agricultural policies and management practices in Egypt and comparable agroecological regions.

References

Abbady, K. A., El-Maaz, E. I. M., Ahmed, H. M. R. M., & Zohry, A. A. (2016). Carbon sequestration as a function of intercropping management practices and different nitrogenous fertilizer types. *Journal of Soil Science and Agriculture Engineering, 7*(8), 565–586.

Abdel-Wahab, T. I., & Abd El-Rahman, R. A. (2016). Response of some soybean cultivars to low light intensity under different intercropping patterns with maize. *International Journal of Applied Agricultural Sciences., 2*(2), 21–31. https://doi.org/10.11648/j.ijaas.20160202.11

Abdel-Wahab, S. I., & El Manzlawy, A. M. (2016). Yield and quality of intercropped sunflower with soybean under different sunflower plant spacings and slow-release nitrogen fertilizer rates in sandy soil. *International Journal of Applied Agricultural Sciences, 2*(3), 32–43.

Abou-Keriasha, M. A., Mohamed, W. M. A., Eiasa, N. M., & Kamel, A. S. (2012a). Intensive crop rotations to improve agricultural production in middle Egypt. *Egyptian Journal of Agricultural Research, 90*(4), 427–443.

Abou-Keriasha, M. A., et al. (2012b). Productivity and soil fertility of legume-based cropping systems in Middle Egypt. *Egyptian Journal of Agronomy, 34*, 15–28.

Abou-Keriasha, M. A., Eisa, N. M. A., & El-Wakil, N. M. H. (2013). Effects of intercropping Faba bean on onion and wheat with or without inoculated bacteria on yields of the three crops. *Egyptian Journal of Agronomy, 35*(2), 169–182.

Akchaya, K., Parasuraman, P., Pandian, K., Vijayakumar, S., Thirukumaran, K., MRAF, M., Rajpoot, S. K., & Choudhary, A. K. (2025a). Boosting resource use efficiency, soil fertility, food security, ecosystem services, and climate resilience with legume intercropping: A review. *Frontiers in Sustainable Food Systems, 9*, 1527256. https://doi.org/10.3389/fsufs.2025.1527256

Akchaya, V., Prakash, M., & Senthilkumar, K. (2025b). Role of legumes in improving soil fertility, structure, and nutrient cycling under diversified cropping systems. *Journal of Soil Science and Plant Nutrition, 25*, 1120–1134.

Arshad, M., Khedher, K. M., Ayed, H., Mouldi, A., Moghanm, F. S., El Ouni, M. H., et al. (2020). Effects of land use and cultivation histories on the distribution of soil organic carbon stocks in the area of the northern Nile Delta in Egypt. *Carbon Management, 11*(4), 341–354. https://doi.org/10.1080/17583004.2020.1790241

Awad, M. M., El-Mehy, A., & Hassan, H. H. (2023). Impact of crop sequence and nitrogen fertilization levels on wheat (*Triticum aestivum L.*) productivity in Egypt. *Egyptian Journal of Agricultural Research, 101*(1), 83–94.

Banerjee, A., Jhariya, M. K., Yadav, D. K., & Raj, A. (2020). *Environmental and sustainable development through forestry and other resources*. Apple Academic Press, CRC Press- a Tayler.

Barbieri, P., Starck, T., Voisin, A. S., & Nesme, T. (2023). Biological nitrogen fixation of legumes crops under organic farming as driven by cropping management: A review. *Agricultural Systems, 205*, 103579. https://doi.org/10.1016/j.agsy.2022.103579

Bardgett, R. D., & van der Putten, W. H. (2014). Belowground biodiversity and ecosystem functioning. *Nature, 515*(7528), 505–511.

Blanco-Canqui, H., Shaver, T. M., Lindquist, J. L., Shapiro, C. A., Elmore, R. W., Francis, C. A., & Hergert, G. W. (2015). Cover crops and ecosystem services: Insights from studies in temperate soils. *Agronomy Journal, 107*, 2449–2474.

Bondok, A. T., & Saad-Allah, K. M. (2024). Optimizing water requirements for green and dry fodder yield in Egyptian clover (Trifolium alexandrinum L.) using sub-surface irrigation in old lands. *International Journal of Agronomy and Agricultural Research, 24*(6), 27–39.

Brooker, R. W., Bennett, A. E., & Cong, W.-F. (2015). Improving intercropping: A synthesis of research in agronomy, plant physiology and ecology. *New Phytologist*. https://doi.org/10.1111/nph.13132

Chia, S. Y., & Lim, M. W. (2022). A critical review on the influence of humidity for plant growth forecasting. *33rd symposium of Malaysian chemical engineers (SOMChE 2022) IOP Conf. Series: Materials science and engineering 1257 012001* IOP Publishing doi:https://doi.org/10.1088/1757-899X/1257/1/012001.

Cong, W. F., Hoffland, E., Li, L., Six, J., Sun, J. H., & Bao, X. G. (2015). Intercropping enhances soil carbon and nitrogen. *Global Change Biology, 21*, 1715–1726. https://doi.org/10.1111/gcb.12738

da Silva, J. P., Teixeira, R., da Silva, I. R., Soares, E. M. P., & Lima, A. M. N. (2021). Decomposition and nutrient release from legume and non-legume residues in a tropical soil. *European Journal of Soil Science*. https://doi.org/10.1111/ejss.13151

Dingha, B. N., Omaliko, P. C., Amoah, B. A., Jackai, L. E., & Shrestha, D. (2021). Evaluation of cowpea (*Vigna unguiculata*) in an intercropping system as pollinator enhancer for increased crop yield. *Sustainability, 13*, 9612. https://doi.org/10.3390/su13179612

Dornbush, M. E., & von Haden, A. C. (2017). Intensified agroecosystems and their effects on soil biodiversity and soil functions. In *Soil health and intensification of agroecosystems* (pp. 173–193). Academic Press. https://doi.org/10.1016/B978-0-12-805317-1.00008-7

Drinkwater, L. E., & Snapp, S. S. (2007). Nutrients in agroecosystems: Rethinking the management paradigm. *Advances in Agronomy, 92*, 163–186.

Drinkwater, L. E., Wagoner, P., & Sarrantonio, M. (1998). Legume-based cropping systems have reduced carbon and nitrogen losses. *Nature, 396*(19), 262–265.

Dutta, A., Trivedi, A., Nath, C. P., Gupta, D. S., & Hazra, K. K. (2022). A comprehensive review on grain legumes as climate-smart crops: Challenges and prospects. *Environmental Challenges, 7*, 100479. https://doi.org/10.1016/j.envc.2022.100479

El-Dessougi, H. A., Abdel-Ghaffar, A. M., & El-Shahawy, T. A. (2018). Phosphorus availability and uptake in calcareous soils as influenced by legume cultivation and rhizosphere processes. *Journal of Plant Nutrition, 41*, 1896–1908.

El-Ghobashy, Y. E., Shams, A. S., & Lamlom, M. M. (2018). Maximizing land use efficiency by intercropping cowpea with some maize cultivars under different maize planting geometries. *Agricultural Sciences, 9*, 1601–1620.

El-Mehy, A. A., Shehata, M. A., Mohamed, A. S., Saleh, S. A., & Suliman, A. A. (2023). Relay intercropping of maize with common dry beans to rationalize nitrogen fertilizer. *Frontiers in Sustainable Food Systems, 7*, 1052392.

Elsadany, M. F. I., & El-Shamy, M. A. (2014). Influence of faba bean-tomato intercropping on mites, insects infestation and the relationship between the pests and predators, leaf phenols content, and yield components of both crops in different intercropping systems. *Mansoura Journal of Biology.*, 55–68.

El-Shamy, M. A., Abdel-Wahab, T. I., Abdel-Wahab, S. I., & Ragheb, S. B. (2015). Advantages of intercropping soybean with maize under two maize plant distributions and three mineral nitrogen fertilizer rates. *Advances in Bioscience and Bioengineering., 3*(4), 30–48. https://doi.org/10.11648/j.abb.20150304.11

FAO. (2003). *On-farm trials for adapting and adopting good agricultural practices*. Food and Agriculture Organization of The United Nations.

FAO. (2013). *Egyptian clover (Trifolium alexandrinum) — Crop guide / technical note* (FAO country/technical note).

FAO. (2019). *Recarbonization of global soils—Egypt country profile*. FAO.

Gaba, S., Lescourret, F., Boudsocq, S., Enjalbert, J., Hinsinger, P., Journet, E. P., Navas, M. L., Wery, J., Louarn, G., Malezieux, E., & Pelzer, E. (2015). Multiple cropping systems as drivers for providing multiple ecosystem services: From concepts to design. *Agronomy for Sustainable Development, 35*, 607–623.

Gazoulis, K. M. I., Kontogeorgou, D. M. V., & Kanatas, P. a. T. I. (2025a). Enhancing ecosystem services in agriculture: The special role of legume intercropping. *Frontiers in Sustainable Food Systems, 9*, 1547879. https://doi.org/10.3389/fsufs.2025.1547879

Gazoulis, I., Kanatas, P., Petraki, D., Antonopoulos, N., Kokkini, M., Danaskos, M., & Travlos, I. (2025b). Enhancing agroecosystem sustainability by means of cover crops in the era of climate change. *Agronomy, 15*, 1051. https://doi.org/10.3390/agronomy15051051

Gebru, H. A. (2015). Review on the comparative advantage of intercropping systems. *Journal of Biology, Agriculture and Healthcare., 6*, 2.

Gomaa, M. A., Zen El-Dein, A. A. M., El-Sorady, G. A., & Salama, N. G. A. (2021). Productivity of maize in relation to preceding crops. *Egyptian Academic Journal of Biological Sciences. H, Botany, 12*, 135–145.

Gontijo, L. M. (2019). Engineering natural enemy shelters to enhance conservation biological control in field crops. *Biological Control, 130*(4), 155–163. https://doi.org/10.1016/j.biocontrol.2018.10.014

Guinet, M., Nicolardot, B., & Voisin, A.-S. (2020). Nitrogen benefits of ten legume pre-crops for wheat assessed by field measurements and modelling. *European Journal of Agronomy, 120*, 126151. https://doi.org/10.1016/j.eja.2020.126151

Gulia, V., Singh Dhull, S., Beniwal, D., Deepali, D., Sangwan, A., & Rani, J. (2025). Legumes and their role in linking nutrition, sustainability and veganism. *International Journal of Vegetable Science, 31*(4), 593–611. https://doi.org/10.1080/19315260.2025.2496739

Hamd-Alla, W. A., Shalaby, E. M., Dawood, R. A., & Zohry, A. A. (2014). Effect of cowpea (Vigan sinensis L.) with maize (Zea mays L.) intercropping on yield and its components. *International Scholarly and Scientific Research and Innovation, 8*(11), 1170–1176.

Hassen, A., Talore, D. G., Tesfamariam, E. H., Friend, M. A., & Mpanza, T. D. E. (2017). Potential use of forage-legume intercropping technologies to adapt to climate change impacts on mixed crop-livestock systems in Africa: A review. *Regional Environmental Change, 17*, 1713–1724. https://doi.org/10.1007/s10113-017-1131-7

Hossain, A., Krupnik, T. J., Timsina, J., Mahboob, M. G., Chaki, A. K., Farooq, M., et al. (2020). Agricultural land degradation: Processes and problems undermining future food security. In *Environment, climate, plant and vegetation growth* (pp. 17–61). Springer International Publishing. https://doi.org/10.1007/978-3-030-49732-3_2

Hossain, M. Z., Anawar, H. M. D., & Chaudhary, D. R. (2023). Legume plants in the context of global climate change: Challenges and scopes for environmental sustainability. In *Climate change and legumes*. Taylor & Francis Books. https://doi.org/10.1201/9781003214885

Huang, J., Tichit, M., Poulot, M., Darly, S., Li, S. C., Petit, C., & Aubry, C. (2015). Comparative review of multifunctionality and ecosystem services in sustainable agriculture. *Journal of Environmental Management, 149*, 138–147.

Hufnagel, J., Reckling, M., & Ewert, F. (2020). Diverse approaches to crop diversification in agricultural research. A review. *Agronomy for Sustainable Development, 40*. https://doi.org/10.1007/s13593-020-00617-4

Jahanzad, E., Barker, A. V., Hashemi, M., Eaton, T., Sadeghpour, A., & Weis, S. A. (2016). Nitrogen release dynamics and decomposition of buried and surface cover crop residues. *Agronomy Journal, 108*, 1735–1741. https://doi.org/10.2134/agronj2016.01.0001

Jensen, E. S., Peoples, M. B., & Hauggaard-Nielsen, H. (2010). Faba bean in cropping systems. *Field Crops Research, 115*, 203–216.

Jensen, E. S., Peoples, M. B., Boddey, R. M., et al. (2012). Legumes for mitigation of climate change and the provision of feedstock for biofuels and biorefineries. A review. *Agronomy for Sustainable Development, 32*, 329–364. https://doi.org/10.1007/s13593-011-0056-7

Jhariya, M. K., Meena, R. S., & Banerjee, A. (2021). Ecological intensification of natural resources towards sustainable productive system. In *Ecological intensification of natural resources for sustainable agriculture*. Springer. https://doi.org/10.1007/978-981-33-4203-3_1

Khan, N., Jhariya, M. K., Raj, A., Banerjee, A., & Meena, R. S. (2021). Eco-designing for sustainability. In M. K. Jhariya, R. S. Meena, & A. Banerjee (Eds.), *Ecological intensification of natural resources for sustainable agriculture*. Springer. https://doi.org/10.1007/978-981-33-4203-3_16

King, A. E., & Hofmockel, K. S. (2017). Diversified cropping systems support greater microbial cycling and retention of carbon and nitrogen. *Agriculture, Ecosystems and Environment, 240*, 66–76.

Klein, A. M., Vaissière, B. E., Cane, J. H., Steffan-Dewenter, I., Cunningham, S. A., Kremen, C., & Tscharntke, T. (2007). Importance of pollinators in changing landscapes for world crops. *Proceedings of the Royal Society B: Biological Sciences, 274*(1608), 303–313.

Kocira, A., Staniak, M., Tomaszewska, M., Kornas, R., Cymerman, J., Panasiewicz, K., & Lipińska, H. (2020). Legume cover crops as one of the elements of strategic weed management and soil quality improvement. A Review. *Agriculture, 10*, 394. https://doi.org/10.3390/agriculture10090394

Kokkini, M., Gazoulis, I., Danaskos, M., Kontogeorgou, V., Kanatas, P., & Travlos, I. (2025). Enhancing ecosystem services in agriculture: The special role of legume intercropping. *Frontiers in Sustainable Food Systems, 9*, 1547879. https://doi.org/10.3389/fsufs.2025.1547879

Krasilnikov, P., Taboada, M. A., & Amanullah. (2022). Fertilizer use, soil health and agricultural sustainability. *Agriculture, 12*, 462. https://doi.org/10.3390/agriculture12040462

Kumar, R., Mishra, V., Buzan, J., Kumar, R., Shindell, D., & Huber, M. (2017). Dominant control of agriculture and irrigation on urban heat island in India. *Scientific Reports, 7*, 14054.

Kumar, S., Meena, R. M., Lal, R., et al. (2018). Role of legumes in soil carbon sequestration. In R. Meena, A. Das, G. Yadav, & R. Lal (Eds.), *Legumes for soil health and sustainable management*. Springer. https://doi.org/10.1007/978-981-13-0253-4_4

Kumar, S., Meena, R. S., & Jhariya, M. K. (2020). *Resources use efficiency in agriculture* (p. 760). Springer. https://doi.org/10.1007/978-981-15-6953-1

Kumari, A., Dash, M., Jagadesh, M., & Singh, S. K. (2025). Impact of climate change on food crops cultivation in the present consequence of global warming. In B. Pramanick, S. V. Singh, S. Maitra, S. Celletti, & A. Hossain (Eds.), *Climate-smart agricultural technologies. Sustainability sciences in Asia and Africa*. Springer. https://doi.org/10.1007/978-981-96-7699-6_1

Lal, R. (2017). Improving soil health and human protein nutrition by pulses-based cropping systems. *Advances in Agronomy, 145*, 167–204. https://doi.org/10.1016/bs.agron.2017.05.003

Lal, R. (2020). Managing soils for negative feedback to climate change and positive impact on food and nutritional security. *Soil Science and Plant Nutrition, 66*, 1–9.

Landschoot, S., Zustovi, R., Dewitte, K., Randall, N. P., Maenhout, S., & Haesaert, G. (2024). Cereal-legume intercropping: A smart review using topic modelling. *Frontiers in Plant Science, 14*, 1228850. https://doi.org/10.3389/fpls.2023.1228850

Lazali, M., & Drevon, J. J. (2023). Legume ecosystemic services in agro-ecosystems: A review. *Communications in Plant Sciences, 2023*. https://doi.org/10.26814/cps2023002

Letourneau, D. K., Armbrecht, I., Rivera, B. S., Lerma, J. M., Carmona, E. J., Daza, M. C., Escobar, S., Galindo, V., Gutiérrez, C., López, S. D., Mejía, J. L., Rangel, A. M. A., Rangel, J. H., Rivera, L., Saavedra, C. A., Torres, A. M., & Trujillo, A. R. (2011). Does plant diversity benefit agroecosystems? A synthetic review. *Ecological Applications, 21*(1), 9–21. https://doi.org/10.1890/09-2026.1

Li, L., Tilman, D., Lambers, H., & Zhang, F. S. (2014). Biodiversity and overyielding: Insights from belowground facilitation of intercropping in agriculture. *Field Crops Research, 165*, 120–127.

Li, C. J., Hoffland, E., Kuyper, T. W., Yu, Y., Zhang, C., Li, H., Zhang, F., & van der Werf, W. (2020). Syndromes of production in intercropping impact yield gains. *Nature Plants, 6*, 653–660. https://doi.org/10.1038/s41477-020-0680-9

Lin, B. B. (2011). Resilience in agriculture through crop diversification: Adaptive management for environmental change. *Agriculture, Ecosystems & Environment, 144*(1), 111–119.

Liu, Q., Sun, X., Wu, W., Liu, Z., Fang, G., & Yang, P. (2022). Agroecosystem services: A review of concepts, indicators, assessment methods and future research perspectives. *Ecological Indicators, 142*, 109218. https://doi.org/10.1016/j.ecolind.2022.109218

Liu, H., Gao, X., Li, C., Cai, Y., Song, X., & Zhao, X. (2025). Intercropping increases plant water availability and water use efficiency: A synthesis. *Agriculture, Ecosystems & Environment, 379*, 109360. https://doi.org/10.1016/j.agee.2024.109360

Lucia, C., Laudicina, V. A., Paliaga, S., Gristina, L., & Muscarella, S. M. (2025). Response of soil microbial biomass and activity to cover crop incorporation methods. *Agronomy, 15*, 2504. https://doi.org/10.3390/agronomy15112504

Lulie, B. (2017). Intercropping practice as an alternative pathway for sustainable agriculture: A review. *ARJASR, 5*, 440–452. https://doi.org/10.14662/ARJASR2017.057

Lüscher, A., Mueller-Harvey, I., Soussana, J. F., Rees, R. M., & Peyraud, J. L. (2014). Potential of legume-based grassland–livestock systems in Europe: A review. *Grass and Forage Science, 69*, 206–228. https://doi.org/10.1111/gfs.12124

Marteau-Bazouni, M., Jeuffroy, M.-H., & Guilpart, N. (2024a). Grain legume response to future climate and adaptation strategies in Europe: A review of simulation studies. *European Journal of Agronomy, 153*, 127056. https://doi.org/10.1016/j.eja.2023.127056

Marteau-Bazouni, M., Moussadek, R., & Mrabet, R. (2024b). Legume-driven shifts in soil microbial communities and implications for nutrient cycling in agroecosystems. *Applied Soil Ecology, 195*, 105230.

Martin-Guay, M., Paquette, A., Dupras, J., & Rivest, D. (2018). The new green revolution: Sustainable intensification of agriculture by intercropping. *Science of the Total Environment, 615*, 767–772.

Masson-Boivin, C., & Sachs, J. L. (2018). Symbiotic nitrogen fixation by rhizobia—The roots of a success story. *Current Opinion in Plant Biology, 44*, 7–15. https://doi.org/10.1016/j.pbi.2017.12.001

Meena, R. S., Kumar, V., Yadav, G. S., & Mitran, T. (2015). Response and interaction of Brady rhizobium japonicum and arbuscular mycorrhizal fungi in the soybean rhizosphere: A review. *Plant Growth Regulation, 84*(2), 207–223.

Metwally, A. A., Safina, S. A., EL-Killany, R., & Saleh, N. E. (2017). Growing corn and soybean in solid and intercropping systems under different levels of irrigation water. *Bioscience Research, 14*(3), 532–541.

Ministry of Agriculture and Land Reclamation. (2024). Desert Research Center Egypt's National Action Plan (NAP) to Combat Desertification, Land Degradation and Drought, 2024–2030. Egypt.

Nanganoa, L. T., Njukeng, J. N., Ngosong, C., Atache, S. K. E., Yinda, G. S., Ebonlo, J. N., et al. (2019). Short-term benefits of grain legume fallow systems on soil fertility and farmers livelihood in the humid forest zone of Cameroon. *International Journal of Sustainable Agricultural Research, 6*, 213–223. https://doi.org/10.18488/journal.70.2019.64.213.223

Nazir, M. J., Li, G., Nazir, M. M., Zulfiqar, F., Siddique, K. H. M., Iqbal, B., & Du, D. (2024). Harnessing soil carbon sequestration to address climate change challenges in agriculture. *Soil and Tillage Research, 237*, 105959. https://doi.org/10.1016/j.still.2023.105959

Nyawade, S. O., Gachene, C. K. K., Karanja, N. N., Gitari, H. I. E., Schulte-Geldermann, & Parker, M. L. (2019). Controlling soil erosion in smallholder potato farming systems using legume intercrops. *Geoderma Regional, 15*, e00225.

Ortiz-Monasterio, I., Wassmann, R., Govaerts, B., Hosen, Y., Katayanagi, N., & Verhulst, N. (2010). Greenhouse gas mitigation in the main cereal systems: Rice wheat and maize. In *Climate change and crop production* (pp. 151–176). CABI.

Ouda, S., & Zohry, A. (2023). Increasing sustainability by intercropping legume crops with sugar beet under imposed water stress. *Irrigation and Drainage*, 1–26. Available from:. https://doi.org/10.1002/ird.2884

Ouda, S., & Zohry, A. (2026). Optimizing resource utilization through intercropping soybeans (Glycine max (L.) Merr.) and maize (Zea mays L.) in Middle Egypt. *Acta Agriculturae Slovenica, 122*(1), 1–23.

Ouda, S., Zohry, A., & Abdel-Wahab, S. (2020). Sustainable intensive cropping to reduce irrigation-induced erosion: II. Changing cropping sequence under sprinkler irrigation practice. *Moroccan Journal of Agricultural Science., 1*(1), 26–33.

Paustian, K., Lehmann, J., Ogle, S., Reay, D., Robertson, G. P., & Smith, P. (2019). Climate-smart soils. *Nature, 532*, 49–57.

Peoples, M. B., Brockwell, J., Herridge, D. F., Rochester, I. J., Alves, B. J. R., Urquiaga, S., Boddey, R. M., Dakora, F. D., Bhattarai, S., Maskey, S. L., et al. (2009). The contributions of nitrogen-fixing crop legumes to the productivity of agricultural systems. *Symbiosis, 48*, 1–17.

Pierre, J. F., Latournerie-Moreno, L., Garruña-Hernández, R., Jacobsen, K. L., Laboski, C. A. M., Salazar-Barrientos, L. D. L., & Ruiz-Sánchez, E. (2021). Farmer perceptions of adopting novel legumes in traditional maize-based farming systems in the Yucatan peninsula. *Sustainability, 13*(20), 11503. https://doi.org/10.3390/su132011503

Pierre, J. F., Jacobsen, K. L., Latournerie-Moreno, L., Torres-Cab, W. J., Chan-Canché, R., & Ruiz-Sánchez, E. (2023). A review of the impact of maize-legume intercrops on the diversity and abundance of entomophagous and phytophagous insects. *PeerJ, 11*, e15640. https://doi.org/10.7717/peerj.15640

Power, A. G. (2010). Ecosystem services and agriculture: Tradeoffs and synergies. *Philosophical Transactions of the Royal Society B: Biological Sciences, 365*, 2959–2971.

Preissel, S., Reckling, M., Schläfke, N., & Zander, P. (2015). Magnitude and farm-economic value of grain legume pre-crop benefits in Europe: A review. *Field Crops Research, 175*, 64–79. https://doi.org/10.1016/j.fcr.2015.01.012

Raffa, D. W., Migliore, M., Campanelli, G., Leteo, F., & Trinchera, A. (2022). Effects of faba bean strip cropping in an outdoor organic tomato system on soil nutrient availability, production, and N budget under different fertilizations. *Agronomy, 12*, 1372. https://doi.org/10.3390/agronomy12061372

Rahman, M., Alam, M. S., Islam, M., Kamal, M. Z. U., Rahman, G. K. M. M., et al. (2022). Potential of legume-based cropping systems for climate change adaptation and mitigation. In *Advances in legumes for sustainable intensification* (pp. 381–402). Academic Press. https://doi.org/10.1016/B978-0-323-85797-0.00030-6

Raj, A., Jhariya, M. K., Khan, N., Banerjee, A., & Meena, R. S. (2021). Ecological intensification for sustainable development. In *Ecological intensification of natural resources for sustainable agriculture*. Springer, , eBook ISBN 978–981–334-203-3. Hardcover ISBN: 978-981-334-202-6. https://doi.org/10.1007/978-981-33-4203-3_5

Rakotomalala, A. A., Ficiciyan, A. M., & Tscharntke, T. (2023). Intercropping enhances beneficial arthropods and controls pests: A systematic review and meta-analysis. *Agriculture, Ecosystems & Environment., 356*, 108617. https://doi.org/10.1016/j.agee.2023.108617

Rakshit, A., Sarkar, B., & Abhilash, P. C. (2018). *Soil amendments for sustainability: Challenges and perspectives*. CRC Press.

Rana, K., Goyal, N., & Sharma, G. P. (2018). Staging stewards of agro-ecosystems in the ecosystem services framework. *Ecosystem Services, 33*, 89–101.

Rodríguez, B. C., Zuazo, V. H. D., Rodríguez, M. S., García-Tejero, I. F., Ruiz, B. G., et al. (2022). Legumes protect the soil erosion and ecosystem services. In *Advances in legumes for sustainable intensification* (pp. 247–266). Academic Press. https://doi.org/10.1016/B978-0-323-85797-0.00010-0

Sainio, P. P., Jauhiainen, L., Honkavaara, E., Wittke, S., Karjalainen, M., & Puttonen, E. (2019). Pre-crop values from satellite images for various previous and subsequent crop combinations. *Frontiers in Plant Science, 10*. https://doi.org/10.3389/fpls.2019.00462

Sarkar, S., Skalicky, M., Hossain, A., Brestic, M., Saha, S., Garai, S., Ray, K., & Brahmachari, K. (2020). Management of crop residues for improving input use efficiency and agricultural sustainability. *Sustainability, 12*, 9808. https://doi.org/10.3390/su12239808

Schnarr, C., Schipanski, M., & Tatarko, J. (2022). Crop residue cover dynamics for wind erosion control in a dryland, no-till system. *Journal of Soil and Water Conservation, 77*(3), 221–229. https://doi.org/10.2489/jswc.2022.00005

Schumacher, M., Dieterich, M., & Gerhards, R. (2020). Effects of weed biodiversity on the ecosystem service of weed seed predation along a farming intensity gradient. *Global Ecology and Conservation, 24*, e01316.

Sheha, A. M., Hamdan, M. K., Abou-Elela, A. M., & Abd Elkader, N. I. (2016). Effect of preceding crop and nitrogen rate on grain yield and yield components of maize. *Agricultural Research Kafr El-Sheikh University, 42*(4), 532–545.

Shumet, S. T., Ayalew, T., Roro, A. G., & Beshir, H. M. (2022). Intercropping and *Rhizobium* inoculation affected microclimate and performance of common bean (*Phaseolus vulgaris L.*) varieties. *Scientifica*, 3471912. https://doi.org/10.1155/2022/3471912

Smitha, G. R., Basak, B. B., Thondaiman, V., & Saha, A. (2019). Nutrient management through organics, bio-fertilizers and crop residues improves growth, yield and quality of sacred basil (Ocimum Sanctum Linn). *Industrial Crops and Products, 128*, 599–606.

Snapp, S. S., Cox, C. M., & Peter, B. G. (2019). Multipurpose legumes for smallholders in sub-Saharan Africa: Identification of promising 'scale out' options. *Global Food Security, 23*, 22–32. https://doi.org/10.1016/j.gfs.2019.03.002

Srivastava, S., Singh, V., & Kumar, R. (2025a). Biological nitrogen fixation and ecosystem services of legumes in sustainable agriculture. *Agronomy for Sustainable Development, 45*, 12.

Srivastava, R. K., Yetgin, A., & Srivastava, S. (2025b). The role of legume roots in carbon sequestration, soil health enhancement, and salinity mitigation under climate change: A comprehensive review. *Soil and Tillage Research, 253*, 106656. https://doi.org/10.1016/j.still.2025.106656

Stagnari, F., Maggio, A., Galieni, A., et al. (2017). Multiple benefits of legumes for agriculture sustainability: An overview. *Chemical and Biological Technologies in Agriculture, 4*(2). https://doi.org/10.1186/s40538-016-0085-1

Stomph, T. J., Dordas, C., & Corre-Hellou, G. (2020a). Nitrogen fixation and transfer in legume-based intercropping systems: A review. *Field Crops Research, 246*, 107681.

Stomph, T., Dordas, C., Baranger, A., de Rijk, J., Dong, B., Evers, J., et al. (2020b). Designing intercrops for high yield, yield stability and efficient use of resources: Are there principles? *Advances in Agronomy, 160*, 1–50. https://doi.org/10.1016/bs.agron.2019.10.002

Vadez, V., Berger, J. D., Warkentin, T., Asseng, S., Ratnakumar, P., Rao, K. P. C., & Sharma, H. C. (2012). Adaptation of grain legumes to climate change: A review. *Agronomy for Sustainable Development, 132*, 31–44. https://doi.org/10.1007/s13593-011-0020-6

Wang, X., Chen, Y., Yang, K., Duan, F., Liu, P., Wang, Z., et al. (2021). Effects of legume intercropping and nitrogen input on net greenhouse gas balances, intensity, carbon footprint and crop productivity in sweet maize cropland in South China. *Journal of Cleaner Production, 314*, 127997.

Weiner, M., Moakes, S., Raya-Sereno, M. D., & Cooper, J. (2024). Legume-based crop rotations as a strategy to mitigate fluctuations in fertilizer prices? A case study on bread wheat genotypes in northern Spain using life cycle and economic assessment. *European Journal of Agronomy, 159*, 127267. https://doi.org/10.1016/j.eja.2024.127267

Wheeler, T., & von Braun, J. (2013). Climate change impacts on global food security. *Science, 341*(6145), 508–513.

Xiong, H., Shen, H., Zhang, L., Zhang, Y., Guo, X., Wang, P., Duan, P., Ji, C., Zhong, L., Zhang, F., & Zuo, Y. (2013). Comparative proteomic analysis for assessment of the ecological significance of maize and peanut intercropping. *Journal of Proteomics, 78*, 447–460.

Yadav, G. S., Lal, R., Meena, R. S., Babu, S., Das, A., Bhomik, S. N., Datta, M., Layak, J., & Saha, P. (2017). Conservation tillage and nutrient management effects on productivity and soil carbon sequestration under double cropping of Rice in north eastern region of India. *Ecology in India*. http://www.sciencedirect.com/science/article/pii/S1470160X17305617

Yang, L., Luo, Y., Lu, B., Zhou, G., Chang, D., Gao, S., et al. (2023). Long-term maize and pea intercropping improved subsoil carbon storage while reduced greenhouse gas emissions. *Agriculture, Ecosystems and Environment, 349*, 108444.

Yu, Y., Stomph, T. J., Makowski, D., Zhang, L., & van der Werf, W. (2016). A meta-analysis of relative crop yields in cereal/ legume mixtures suggests options for management. *Field Crops Research, 198*, 269–279.

Yuvaraj, M., Kumar, S., & Sharma, P. (2020a). Soil microbial diversity and nutrient cycling under legume-based cropping systems. *Soil Biology & Biochemistry, 148*, 107906.

Yuvaraj, M., Pandiyan, M., & Gayathri, B. (2020b). Role of legumes in improving soil fertility status. In *Legume crops—Prospects, production and uses*. Intechopen. https://doi.org/10.5772/intechopen.93247

Zabala, J. A., Martínez-Paz, J. M., & Alcon, F. (2021). Comprehensive approach for agroecosystem services and disservices valuation. *Science of the Total Environment, 768*, 1–14.

Zang, H., Yang, X., Feng, X., Qian, X., Hu, Y., Ren, C., & Zeng, Z. (2015). Rhizodeposition of nitrogen and carbon by mungbean (Vigna radiata L.) and its contribution to intercropped oats (Avena nuda L.). *PLoS One, 10*(3), e0121132. https://doi.org/10.1371/journal.pone.0121132

Zen El-Dein, A. A. M., Koriem, M. H. M., Alsubeie, M. S., Alsalmi, R. A., Masrahi, A. S., Al-Harbi, N. A., Al-Qahtani, S. M., Awad-Allah, M. M. A., & Hefny, Y. A. A. (2022). Effect of mycorrhiza fungi, preceding crops, mineral and bio fertilizers on maize intercropping with cowpea. *Agriculture, 12*, 1934. https://doi.org/10.3390/agriculture12111934

Zhang, W., Ricketts, T. H., Kremen, C., Carney, K., & Swinton, S. M. (2007). Ecosystem services and dis-services to agriculture. *Ecological Economics, 64*, 253–260.

Zohry, A., & Ouda, S. (2025). Legume crops enhance water use efficiency under intercropping system with wheat. *Moroccan Journal of Agricultural Sciences, 6*(3), 145–160.

Zohry, A. A., Ouda, S., Hamd-Alla, W., & Shalaby, E. (2017). Evaluation of different crop sequences for wheat and maize in sandy soil. *Acta Agriculturae Slovenica., 109*(2), 383–392.

Zohry, A., Ouda, S., & Abdel-Wahab, T. (2020). Sustainable intensive cropping to reduce irrigation-induced erosion: I. Intercropping systems under surface irrigation practice. *Moroccan Journal of Agricultural Science, 1*(2), 63–71.

Part I
Regulating Services of Legumes

Chapter 2
Legumes Integration Strategies for Optimizing Soil Carbon Sequestration

Abstract Carbon sequestration is a vital regulatory service within the agroecosystem. Carbon sequestration in agricultural soils is a crucial approach for alleviating climate change while improving soil fertility and sustaining long-term productivity. This chapter analyzes the enhancement of soil organic carbon stocks and carbon sequestration in agroecosystems through the incorporation of legumes into cropping systems via intercropping, preceding non-legumes, and utilizing legumes as cover crops, supported by a comprehensive literature review. Legumes assimilate atmospheric nitrogen through symbiotic relationships with rhizobia, enhancing biomass output, improving soil structure, and promoting microbial activity. Intercropping legumes with cereals or non-cereal crops increases soil organic matter, while legumes precede other crops boost nutrient cycling and stabilize soil organic carbon. Employing legumes as cover crops, especially varieties like fahl clover, enhances soil organic carbon by augmenting organic residues and nitrogen, so boosting aggregate stability and minimizing mineralization losses. These activities collectively enhance soil health, bolster resilience to climate stress, and ensure the long-term viability of agricultural systems.

Keywords Intercropping with legumes · Legumes preceding non-legumes · Utilizing legumes as cover crops · Climate change · Carbon sequestration

Introduction

Agroecosystems are anthropogenically managed ecosystems that deliver vital services for food production, yet encounter significant difficulties like soil degradation, biodiversity decline, and climate variability (Banerjee et al., 2020). A multitude of studies has examined agroecosystem services and investigated methods to enhance their utilization, encompassing the comprehensive assessment of agroecosystem services, the characterization of trade-offs and causal factors, the evaluation of scale effects on ecosystem services, and the optimization strategies for trade-offs in agroecosystem services (Verhagen et al., 2018). Agroecosystem services encompass

A. E. Zohry, S. A. H. Ouda, *The Agroecosystem Services of Leguminous Crops*,
https://doi.org/10.1007/978-3-032-32461-0_2

regulating services and supporting services (Power, 2010). The regulating services of agroecosystems encompass carbon sequestration, climate regulation, waste decomposition, and soil conservation (Liu et al., 2022).

Carbon sequestration is a crucial regulatory service of the agroecosystem. Carbon sequestration refers to the capture and secure storage of carbon that would otherwise be released into the atmosphere, with the objective of alleviating global climate change (Oh, 2010). Carbon sequestration in agriculture encompasses measures that collect atmospheric carbon dioxide and sequester it in soil and plant biomass, serving as a crucial mechanism for mitigating climate change and enhancing soil productivity (Mattila & Vihanto, 2024). It denotes the sequestration of atmospheric CO_2 by plants, which is subsequently deposited in soil organic matter (Lal, 2008). Soil organic matter generally comprises 58% carbon, commonly termed soil organic carbon (Stockmann et al., 2013). This process, known as carbon farming, seeks to mitigate greenhouse gas emissions and enhance soil health, thereby fostering more sustainable agricultural systems (Paustian et al., 2016). Enhancing plant growth conditions can concurrently augment soil carbon sequestration, both directly through increased biomass carbon input and indirectly by fostering root development and microbial activity (Colombi et al., 2019). Augmenting the organic carbon content in soil via improved carbon sequestration can elevate soil microbial biomass (Fierer, 2017) and enhance the complexity of soil community networks, encompassing keystone microbes and mycorrhizal species that are crucial for soil health, plant vitality, and carbon cycling (Xue et al., 2020).

Numerous factors affect the capacity for soil carbon sequestration and its variability, including historical and contemporary management methods, soil properties, climate, environmental conditions, resource and nutrient availability, and microbial decomposition (Zomer et al., 2017). Soil management strategies identified by long-term research can enhance carbon sequestration. Enhancing rotations to augment biomass and organic material inputs into soils, as well as diversifying cropping systems by incorporating legume-based systems, including intercropping, preceding non-legumes, and leguminous cover crops is considered as beneficial management techniques (Zhang et al., 2024).

Legumes offer essential agroecosystem functions by fixing atmospheric nitrogen, hence augmenting soil fertility and crop yield, as well as enhancing soil structure and water retention (Rani et al., 2022). Moreover, legumes enhance biodiversity, promote beneficial insects, and diminish the necessity for synthetic fertilizers, pesticides, and fossil energy inputs, so rendering agricultural systems more sustainable and resilient (Cagasan & Amarado, 2023). Furthermore, the symbiotic association of legumes with nitrogen-fixing Rhizopus in their root nodules renders them exceptional crops for enhancing soil organic matter (Nazir et al., 2024). Legumes enhance carbon sequestration in the soil via their extensive root systems, in addition to nitrogen-fixing capabilities, which promote the accumulation of soil organic matter, functioning as a carbon sink (Hossain et al., 2023). In a three-year crop rotation study by Abou-Keriasha et al. (2012), comprising 50% legume crops, the soil organic matter level rose by 38%. Reports indicate that following the harvest of legume crops, a significant quantity of residual nitrogen is available for subsequent crops, contributing to the enrichment of soil fertility through organic matter (Banik et al., 2018).

Climate change substantially diminishes carbon sequestration due to alterations in temperature, precipitation, and other environmental variables, hence reducing the efficacy of natural carbon sinks such as forests and oceans (Zhen et al., 2023). Climate change can impede photosynthesis, the mechanism by which plants transform CO_2 into energy and biomass (Don et al., 2023). Variations in temperature and water availability may constrain photosynthetic rates, hence diminishing carbon uptake (Enríquez-de-Salamanca, 2024). Climate change could result in soil deterioration, as climate is the primary factor influencing soil formation (Karmakar et al., 2016). Hamidov et al. (2018) assert that climate change can affect soil functionality by disrupting organic carbon transformations and nitrogen cycling, since altered moisture and soil temperature result in less carbon storage. Agricultural land, forests, and wetlands have emerged as significant land-use categories capable of sequestering extra carbon in soils and biomass, therefore reducing atmospheric CO_2 levels and contributing to climate change mitigation (Griscom et al., 2017).

This chapter analyzes the enhancement of soil organic carbon stocks and carbon sequestration in agroecosystems through the incorporation of legumes into cropping systems via intercropping, preceding non-legumes, and employing legumes as cover crops, supported by a comprehensive literature review.

The Function of Legumes in Soil Carbon Sequestration

Incorporating legume crops into the cropping system has been shown to enhance soil quality, as legumes can sequester 30% more soil organic carbon than other species (Kumar et al., 2018). Enhancing soil organic matter typically results in augmented carbon sequestration (Nazir et al., 2024). Soil organic matter predominantly consists of carbon, and its augmentation leads to enhanced carbon sequestration in the soil, hence reducing atmospheric carbon and alleviating climate change (Magnolo et al., 2021). Roohi et al. (2022) discovered that the integration of leguminous crops into the cropping system resulted in improved aggregation and carbon buildup, attributed to increased carbon input from root residues and root length density. The integration of legumes in the cropping systems to enhance soil carbon sequestration can be done through intercropping, legumes preceding non-legumes, and legumes cover crops (Fig. 2.1).

Intercropping Legumes with Non-leguminous Crops

Intercropping legumes can substantially improve carbon sequestration in agricultural systems by augmenting the carbon deposited in the soil (Cong et al., 2015a). Legumes' capacity to fix atmospheric nitrogen enhances biomass output and carbon incorporation into the soil (Lüscher et al., 2014). The robust root systems of legumes enhance soil structure and facilitate soil aggregation (Duarte et al., 2024).

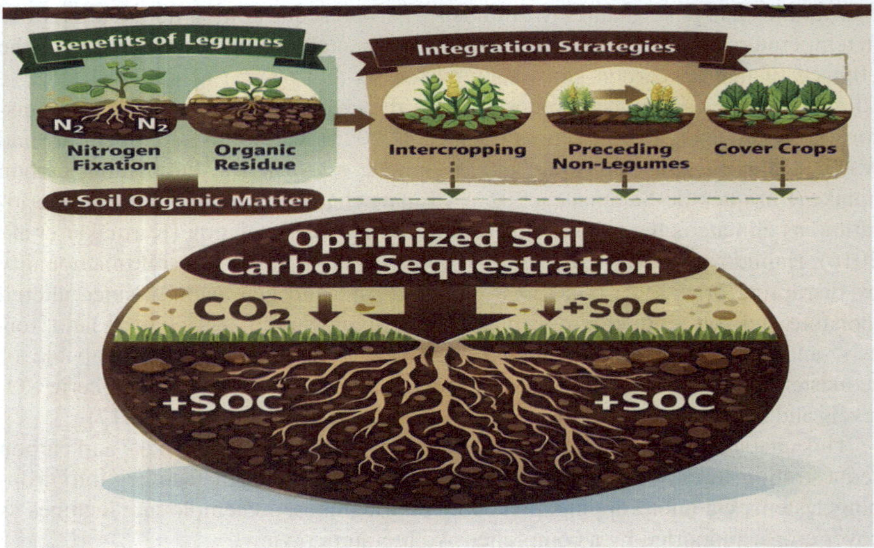

Fig. 2.1 The function of legumes in soil carbon sequestration

Research indicates that intercropping systems contribute more carbon to the soil via root remains than monoculture cropping systems (Li et al., 2011). Moreover, legumes can enhance soil health and microbial activity, hence improving carbon stability (Mukhopadhyay et al., 2021). Cong et al. (2015a) assert that intercropping possesses significant soil carbon sequestration potential, enhancing several agro-ecosystem services through increased production, improved soil quality, and enhanced soil carbon sequestration. Moreover, a modified rhizosphere environment that enhances microbial activity can be attained (Tian et al., 2021). In a long-term field experiment extending over seven consecutive years, Cong et al. (2015b) reported that intercropping systems led to measurable improvements in soil organic carbon dynamics compared with monoculture systems. Specifically, the soil organic carbon content within the upper 0–20 cm soil layer was found to be approximately 4% higher under intercropping. This increase reflects the cumulative effects of diversified plant inputs, including greater root biomass, enhanced rhizodeposition, and more heterogeneous crop residues returned to the soil. Over time, these factors contribute to improved carbon sequestration and stabilization processes in the topsoil.

Intercropping Systems of Legumes with Cereals

Intercropping with leguminous crops provides numerous advantages to the agroecosystem, as legumes enhance soil nitrogen levels through their capacity to fix atmospheric nitrogen via a symbiotic association with soil rhizobia, thereby promoting soil organic carbon sequestration (Liu et al., 2022). Furthermore, Tang et al. (2021) reported that intercropping legumes with cereals resulted in a 20% increase in soil organic matter, thereby improving soil fertility and nutrient availability. Although certain studies indicate a marginal rise in overall soil carbon content, the principal advantage frequently lies in an elevated nitrogen content within the soil organic matter, resulting in a reduction of the carbon-to-nitrogen (C/N) ratio (Cong et al., 2015b).

In this context, intercropping faba bean with wheat might beneficially affect soil organic matter by augmenting soil nitrogen levels and possibly accelerating its decomposition rate, resulting in more productive and resilient soil over time (Kocira et al., 2020). This is due to the nitrogen-fixing capabilities of faba beans, which enhance soil nitrogen content (Zohry & Ouda, 2025). Intercropping faba bean with wheat has been shown to enhance soil organic matter by 31% and improve soil structure (Zhang et al., 2025). Furthermore, Bichel et al. (2016) indicated that intercropping soybean with maize at varying ratios resulted in a greater accumulation of soil organic carbon compared to sole maize, with the highest potential for enhancing soil organic carbon stocks observed in the 2 maize: 3 soybean intercropping system. Roohi et al. (2022) assert that intercropping cowpea with maize enhances carbon sequestration relative to monocultures of each crop.

Legumes–Cereals Intercropping in Egypt

In Egypt, intercropping faba bean with wheat can augment soil organic matter by increasing plant biomass, especially from the leguminous faba bean, which supplies nitrogen and other nutrients via root nodules and residues, resulting in enhanced soil nitrogen stock and improved dynamics of soil organic matter (Abou-Keriasha & Eissa, 2014; Hamada & Hamd-Alla, 2019). Furthermore, Eissa et al. (2017) intercropped three faba bean planting densities, namely 20%, 30%, and 40% of its recommended density with wheat (100% of its recommended planting density), which resulted in increases in soil organic matter of 15%, 18%, and 20%, respectively. Whereas, Abbady et al. (2016) conducted a three-year study, in which pea was intercropped with wheat in the winter season at two densities: 100% peas +50% wheat of its recommended density, and 100% peas +70% wheat of its recommended density. An increase in soil carbon sequestration of 135% was seen when 100% peas were intercropped with 50% wheat. Moreover, intercropping 100% peas with 70% wheat resulted in a 123% increase in soil carbon sequestration.

Intercropping Legumes with Other Crops

It has been reported that intercropping soybean with potato can substantially improve both the quantity and quality of soil organic carbon. In a recent study, Wang et al. (2025) found that this intercropping system increased soil organic carbon content by approximately 8% while enhancing carbon stability by 23%

compared with monocropping systems. These improvements are likely attributed to the complementary interactions between the two crops, including greater root biomass, diverse root exudates, and increased return of organic residues to the soil. These findings implied that the soybean–potato system may stimulate microbial diversity and activity, promoting more efficient carbon cycling and aggregation processes. Collectively, these effects highlight the potential of legume-based intercropping systems to strengthen soil health, support sustainable productivity, and contribute to climate change mitigation strategies.

Legumes–Non-cereals Intercropping in Egypt

Furthermore, in Egypt, Abbady et al. (2016) conducted a three-year study in the summer seasons involving intercropping legumes with non-cereal crops. Cowpea and sunflower were intercropped at two planting densities, namely 50% cowpea +100% sunflower of its recommended density and 70% cowpea +100% sunflower of its recommended density. They reported that carbon sequestration in the soil after three summer seasons rose by 31% for a 50% cowpea +100% sunflower intercropping system, compared to the value obtained from sole sunflower cultivation. An increase of 34% in carbon sequestration was seen in the intercropping system of 70% cowpea +100% sunflower compared to sole sunflower cultivation.

Alternatively, Zohry et al. (2020) conducted a comparative study on the effects of two agricultural management packages on soil organic matter content. The first package involved traditional farming practices that utilized excessive irrigation water and fertilizer for monoculture cultivation of wheat in the winter season and monoculture cultivation of sunflower in the summer season, which is the common farmer practice in the region. The second package featured improved management practices, which included the precise application of irrigation water and fertilizer, to peas intercropped with wheat (50% peas +100% wheat of the recommended density) and cowpea intercropped with sunflower (50% cowpea +100% sunflower of the recommended density). The results showed an increase in soil organic matter by 9% for peas intercropping system with wheat and by 11% for cowpea intercropping system with sunflower following the fourth growing season under enhanced management practices, compared to the values obtained from farmer practice.

In another study by Ouda and Zohry (2024a), fahl clover (a variety of Egyptian clover) was intercropped with wheat and the enhancement in soil organic matter post-harvest was assessed. They reported an increase in soil organic matter by 3% after harvest of fahl clover intercropping system with wheat (35% fahl clover +100% wheat of its recommended densities), compared to the value obtained from a monoculture of wheat.

Leguminous Crops Preceding Non-leguminous Crops

Legumes precede non-legumes augmenting carbon sequestration by improving soil substrate quality, enhancing microbial nutrient availability, and promoting the growth and activity of beneficial microbes, resulting in more stable soil organic carbon and improved microbial nutrient use efficiency (Hu et al., 2023a, b). Legumes fix atmospheric nitrogen, alleviating nitrogen constraints for microorganisms and enabling them to more efficiently convert and stabilize carbon sources, such as legume root biomass and root exudates (Kokkini et al., 2025). Legumes, when grown as preceding crops, increase soil organic matter, hence improving soil structure, water retention, and overall soil health (Stagnari et al., 2017). The incorporation of legumes as a sustainable strategy can enhance crop yields by improving soil multifunctionality and preserving soil organic carbon stores, especially in low-yielding agroecosystems (Liu et al., 2023).

According to Etesami and Adl (2020), when used as a preceding crop, legume roots significantly enhance carbon sequestration through multiple mechanisms that promote soil organic carbon retention, including root biomass inputs (direct carbon addition via root production), the release of root exudates, and interactions with soil microorganisms—particularly rhizobia through rhizodeposition—which collectively support the stabilization of organic carbon in soils. Furthermore, the introduction of crops with elevated C:N residues enhances the carbon and nitrogen content in the upper 20 cm of the soil (Cleveland et al., 2014). Gan et al. (2015) asserted that including legumes into cropping systems can effectively replace summer fallowing in rain-fed arid regions to reduce soil erosion. Panagea et al. (2025) indicated that augmenting the quantity of forage legumes in rotation with other crops resulted in a soil organic accumulation of up to 13.25 Mg/ha (0.44 Mg/ha/year).

In a legumes preceding wheat cropping experiment, the preceding of peas demonstrated significantly enhanced rhizobial nitrogen fixation compared to lentils and chickpeas preceding wheat (Lengwati et al., 2020). Moreover, in the Northern Great Plains of North America, the advantages in wheat yield attributable to previous legumes cultivation fluctuated significantly, reaching up to 35% (Gan et al., 2015). Cultivating legumes as a preceding crop not only increases soil nitrogen through biological nitrogen fixation and the return of crop residues but also improves soil carbon stocks due to the input of high-quality residues (Lal, 2017).

The legume-based rotation system enhanced soil organic carbon and nitrogen in soil aggregates by 9% and 12% after six years, respectively (Zhang et al., 2019). Zuber et al. (2018) assessed the impact of four crop rotations on soil organic carbon concentration. The crop rotations included continuous corn, corn-soybean, corn-soybean-wheat, and continuous soybean. They concluded that corn-soybean-wheat rotation was proven to enhance soil organic carbon relative to alternative rotations.

Leguminous Preceding Crops in Egypt

In Egypt, Abou-Keriasha and Eissa (2014) determined that the inclusion of faba bean and fahl clover as preceding crops to wheat resulted in a 12% increase in soil organic matter. Koriem and Hamd-Alla (2025) conducted a study on different crop sequences, revealing that soil organic matter increased by 48% when soybean preceded sugar beet and by 63% when cowpea preceded sugar beet, in comparison to the values recorded when maize preceded sugar beet. Whereas, Abd El-Hadi et al. (2000) reported that the inclusion of Egyptian clover prior to rice in a three-year crop rotation on salt-affected soil resulted in an increase in soil organic matter by the conclusion of the three years. According to Abd El-Hadi et al. (2002), five two-year crop rotations were conducted in North Egypt under rain-fed conditions to evaluate the accumulation of soil organic matter resulting from these rotations. The rotations included barley-fallow, wheat-wheat, barley-barley, barley-pea, and barley-lentil. The findings revealed that the maximum soil organic matter was achieved through the implementation of a barley-pea rotation, resulting in a 47% increase in organic matter post-harvest. They also reported that the values of N, P, and K were augmented by 73%, 300%, and 19%, respectively, under barley-pea rotation.

Furthermore, Khadr et al. (2004) implemented a crop rotation on sandy soil characterized by a low organic matter content (0.25%) and limited accessible nitrogen (9.25 ppm). They reported that increasing the number of legume crops in the rotation enhances organic matter and macronutrients. They reported an increase in the soil organic matter 32% after 3 years. Furthermore, soil available nitrogen, phosphorus, and potassium were increased by 6%, 46%, and 20%, respectively. According to Abou-Keriasha et al. (2012), a three-year crop rotation was implemented to augment the quantity of farmed legumes as preceding crops to non-legumes, resulting in a 38% increase in soil organic matter.

Employing Legumes as Cover Crops

Cover crops are cultivated between primary crops to enhance soil health and safeguard the land from erosion during the off-season (Akchaya et al., 2025). Common winter legume cover crops in the United States include hairy vetch, red clover, and winter pea (Thapa et al., 2025). Cover crops are defined as non-cash crops grown before or alongside primary crops to sustain soil covered with vegetation for prolonged durations (Blanco-Canqui et al., 2015). These crops also augment carbon levels in soils due to their high carbon content (Han et al., 2016), through photosynthesis, root exudation, and litter decomposition (Blanco-Canqui & Ruis, 2020), while concurrently diminishing soil organic carbon mineralization by regulating soil temperature and enhancing soil aggregate stability (Restovich et al., 2022). Bolinder et al. (2020) asserts that cover crops can enhance soil organic carbon stock by 9–10%. Cover crops are particularly significant as they enhance soil health and bolster resilience against climate change (Ball et al., 2020). Prior research indicates that legume cover crops have a superior capacity compared to non-legume cover crops for enhancing soil organic carbon (Hu et al., 2023a, b). The latest elucidation ascribes this phenomenon to elevated microbial carbon consumption efficiency (Jia et al., 2022). Leguminous cover crops can also be employed to diminish weeds, avert soil erosion, and augment nitrogen levels in the soil for subsequent crops.

Use of Leguminous Cover Crops in Egypt

In Egypt, fahl clover is used as a cover crop after the summer season. Fahl clover is a variety of Egyptian clover characterized by its capacity for stem branching and is harvested only once. This crop is distinguished by its rapid growth and substantial forage yield, capable of being cultivated as an early winter season crop growing from September to early November, with a growing season of 60–70 days (Bakheit et al., 2016). It notably enhances soil quality by enriching it with organic matter and nitrogen through atmospheric nitrogen fixation and the incorporation of organic residues. Its cultivation enhances the physical, chemical, and biological attributes of soil, including pH, water availability, and nutrient concentrations, particularly in newly reclaimed areas (Abbas et al., 2013). Therefore, it may be regarded as an improved fallow. Improved fallow entails cultivating a beneficial legume crop during the fallow phase to rejuvenate soil fertility and augment eventual crop yields (Lemage & Tsegaye, 2020). Fahl clover, as an enhanced fallow, has been reported to augment soil organic matter (Ouda & Zohry, 2024b). According to Abou-Keriasha and Eissa (2014), an increase by 7% in soil organic matter was obtained with the incorporation of fahl clover as an enhanced fallow.

Conclusion

Legume-based cropping strategies—whether intercropped, rotated, or utilized as cover crops—are excellent methods for enhancing soil carbon sequestration and organic matter levels. These systems augment nitrogen availability, boost microbial interactions, and stabilize organic carbon, therefore fostering increased production and soil resilience in the face of climate change. Implementing various legume-rich systems constitutes a pragmatic and sustainable approach to enhancing agricultural productivity and agroecosystem.

References

Abbady, K. A., El-Maaz, E. I. M., Ahmed, H. M. R. M., & Zohry, A. A. (2016). Carbon sequestration as a function of intercropping management practices and different nitrogenous fertilizer types. *Journal of Soil Science and Agriculture Engineering, 7*(8), 565–586.

Abbas, Z. M., Khalil, F. A. F., & Eletr, W. M. T. (2013). Influence of water regimes and soil conditioners on yield, yield components and water utilization of Egyptian clover (variety Fahl). *Journal of Plant Production, 4*(11), 1675–1690.

Abd El-Hadi, A. H., Abo El-Enein, R. A., & Omran, A. (2000). Sustainability of soil fertility status after three-year crop rotation at Middle Egypt region. In *Xth international colloquium for the optimization of plant nutrition. April 8–13, Cairo, Egypt.*

Abd El-Hadi, A. H., Abd El-Shafy, A. M., Aboelenin, R., Haddad, N., & Salh, A. (2002). Soil fertility status as affected by different fertilizer levels and five crop rotations under rain fed conditions at North Sinai. *Egyptian Journal of Soil Science, 42*(3), 463–477.

Abou-Keriasha, M. A., & Eissa, N. M. A. (2014). Mineral nitrogen fertilizer and compost effects on wheat and maize yields in crop sequences including intercropped legumes. *Egyptian Journal of Agronomy, 36*(2), 147–163.

Abou-Keriasha, M. A., Mohamed, W. M. A., Eiasa, N. M., & Kamel, A. S. (2012). Intensive crop rotations to improve agricultural production in Middle Egypt. *Egyptian Journal of Agricultural Research, 90*(4), 427–443.

Akchaya, K., Parasuraman, P., Pandian, K., Vijayakumar, S., Thirukumaran, K., Mustaffa, M. R. A. F., Rajpoot, S. K., & Choudhary, A. K. (2025). Boosting resource use efficiency, soil fertility, food security, ecosystem services, and climate resilience with legume intercropping: A review. *Frontiers in Sustainable Food Systems, 9*, 1527256. https://doi.org/10.3389/fsufs.2025.1527256

Bakheit, B. R., Abo-Elwafa, A., Abdel-Galil, M. M., & Abdelmonem, A. M. A. (2016). Impacts of recurrent selection and synthetic population on forage and seed yields of mono-cut Egyptian Clover (Trifolium alexandrinum L.). *Assiut The Journal of Agricultural Science, 47*(1), 31–44.

Ball, K. R., Baldock, J. A., Penfold, C., Power, S. A., Woodin, S. J., Smith, P., & Pendall, E. (2020). Soil organic carbon and nitrogen pools are increased by mixed grass and legume cover crops in vineyard agroecosystems: Detecting short-term management effects using infrared spectroscopy. *Geoderma*, 114619. https://doi.org/10.1016/j.geoderma.2020.114619

Banerjee, A., Jhariya, M. K., Yadav, D. K., & Raj, A. (2020). *Environmental and sustainable development through forestry and other resources*. Apple Academic Press, CRC Press- a Tayler.

Banik, D., Upadhyay, H., Devi, B. P., & Darvhankar, M. (2018). Effect of different irrigation system on growth and development of wheat-chickpea intercropping. *Bulletin of Environmental Pharmacology and Life Sciences, 8*(1), 32–34.

Bichel, A., Oelbermann, M., Voroney, P., & Echarte, L. (2016). Sequestration of native soil organic carbon and residue carbon in complex agroecosystems. *Carbon Management, 7*, 1–10.

Blanco-Canqui, H., & Ruis, S. J. (2020). Cover crop impacts on soil physical properties: A review. *Soil Science Society of America Journal, 84*, 1527–1576. https://doi.org/10.1002/saj2.20129

Blanco-Canqui, H., Shaver, T. M., Lindquist, J. L., Shapiro, C. A., Elmore, R. W., Francis, C. A., & Hergert, G. W. (2015). Cover crops and ecosystem services: Insights from studies in temperate soils. *Agronomy Journal, 107*, 2449–2474.

Bolinder, M. A., Crotty, F., Elsen, A., et al. (2020). The effect of crop residues, cover crops, manures and nitrogen fertilization on soil organic carbon changes in agroecosystems: A synthesis of reviews. *Mitigation and Adaptation Strategies for Global Change, 25*, 929–952. https://doi.org/10.1007/s11027-020-09916-3

Cagasan, U., & Amarado, J. (2023). Influence of intercropping on the growth and yield performance of upland Rice (Oryza sativa L. var. Zambales) and Peanut (Arachis hypogaea L.). *Research in Agricultural Sciences, 54*(1), 2–8.

Cleveland, C. C., Reed, S. C., Keller, A. B., Nemergut, D. R., et al. (2014). Litter quality versus soil microbial community controls over decomposition: A quantitative analysis. *Oecologia, 174*, 283–294. https://doi.org/10.1007/s00442-013-2758-9

Colombi, T., Walder, F., Büchi, L., Sommer, M., Liu, K., Six, J., van der Heijden, M. G. A., Charles, R., & Keller, T. (2019). On-farm study reveals positive relationship between gas transport capacity and organic carbon content in arable soil. *The Soil, 5*, 91–105.

Cong, W. F., Hoffland, E., Li, L., Six, J., Sun, J. H., & Bao, X. G. (2015a). Intercropping enhances soil carbon and nitrogen. *Global Change Biology, 21*, 1715–1726. https://doi.org/10.1111/gcb.12738

Cong, W. F., Hoffland, E., Li, L., et al. (2015b). Intercropping affects the rate of decomposition of soil organic matter and root litter. *Plant and Soil, 391*, 399–411. https://doi.org/10.1007/s11104-015-2433-5

Don, A., Seidel, F., Leifeld, J., & Kätterer, T. (2023). Carbon sequestration in soils and climate change mitigation—Definitions and pitfalls. *Global Change Biology, 30*, e16983. https://doi.org/10.1111/gcb.16983

Duarte, R. D. C., Iannetta, P. P. M., & Gomes, A. M. a. V. M. W. (2024). More than a meator synthetic nitrogen fertiliser-substitute: A review of legume phytochemicals as drivers of 'one health' via their influence on the functional diversity of soil- and gut-microbes. *Frontiers in Plant Science, 15*, 1337653. https://doi.org/10.3389/fpls.2024.1337653

Eissa, N. M. A., Rashwan, B. R. A., & Abozaed, S. H. F. (2017). Influences of intercropping two legume crops and mineral nitrogen fertilizer on wheat yield and micro-organisms activity in the rhizosphere soil. *Bulletin of the Faculty of Agriculture, Cairo University, 68*, 59–77.

Enríquez-de-Salamanca, Á. (2024). Influence of climate change on carbon sequestration in pine forests of Central Spain. *Atmosphere, 15*, 1178. https://doi.org/10.3390/atmos15101178

Etesami, H., & Adl, S. M. (2020). Can interaction between silicon and non–rhizobial bacteria help in improving nodulation and nitrogen fixation in salinity–stressed legumes? A review. *Rhizosphere, 15*, 100229. https://doi.org/10.1016/j.rhisph.2020.100229

Fierer, N. (2017). Embracing the unknown: Disentangling the complexities of the soil microbiome. *Nature Reviews. Microbiology, 15*, 579–590. https://doi.org/10.1038/nrmicro.2017.87

Gan, Y., Hamel, C., O'Donovan, J. T., Cutforth, H., Zentner, R. P., Campbell, C. A., Niu, Y., & Poppy, L. (2015). Diversifying crop rotations with pulses enhances system productivity. *Scientific Reports, 5*, 14625. https://doi.org/10.1038/srep14625

Griscom, B. W., Adams, J., Ellis, P. W., Houghton, R. A., Lomax, G., Miteva, D. A., Schlesinger, W. H., Shoch, D., Siikamäki, J. V., Smith, P., Woodbury, P., Zganjar, C., Blackman, A., Campari, J., Conant, R. T., Delgado, C., Elias, P., Gopalakrishna, T., Hamsik, M. R., et al. (2017). Natural climate solutions. *Proceedings of the National Academy of Sciences of the United States of America, 114*(44), 11645–11650.

Hamada, A., & Hamd-Alla, W. A. (2019). Productivity of intercropped wheat with faba bean under crop sequences and foliar application of humic acid. *Egyptian Journal of Agronomy, 41*(3), 225–241.

Hamidov, A., Helming, K., Bellocchi, G., Bojar, W., et al. (2018). Impacts of climate change adaptation options on soil functions: A review of European case-studies. *Land Degradation and Development, 29*(8), 2378–2389. https://doi.org/10.1002/ldr.3006

Han, P., Zhang, W., Wang, G., et al. (2016). Changes in soil organic carbon in croplands subjected to fertilizer management: A global meta-analysis. *Scientific Reports, 6*, 27199. https://doi.org/10.1038/srep27199

Hossain, M. Z., Anawar, H. M. D., & Chaudhary, D. R. (2023). Legume plants in the context of global climate change: Challenges and scopes for environmental sustainability. In *Climate change and legumes*. Taylor & Francis Books. https://doi.org/10.1201/9781003214885

Hu, Q., Jiang, T., Thomas, B. W., Chen, J., Xie, J., Hu, Y., et al. (2023a). Legume cover crops enhance soil organic carbon via microbial necromass in orchard alleyways. *Soil and Tillage Research*, 105858. https://doi.org/10.1016/j.still.2023.105858

Hu, Q., Thomas, B. W., Powlson, D., Hu, Y., Zhang, Y., Jun, X., Shi, X., & Zhang, Y. (2023b). Soil organic carbon fractions in response to soil, environmental and agronomic factors under cover cropping systems: A global meta-analysis, agriculture. *Ecosystems & Environment, 355*, 108591. https://doi.org/10.1016/j.agee.2023.108591

Jia, B., Jia, L., Zhang, Y., Mou, X. M., & Li, X. G. (2022). Leguminous Caragana korshinskii evidently enhances microbial necromass carbon accumulation in dryland soils. *Catena*, 106342. https://doi.org/10.1016/j.catena.2022.106342

Karmakar, R., Das, I., Dutta, D., & Rakshit, A. (2016). Potential effects of climate change on soil properties: A review. *Science International, 4*, 51–73.

Khadr, M. S., Abou-El-Enwin, R., El-Shafy, A., Zahran, F., & Zohry, A. (2004). Sustainability of soil fertility status after 3-year crop rotation in sandy soils in Egypt. *Egyptian Journal of Agricultural Research, 82*(2), 475–491.

Kocira, A., Staniak, M., Tomaszewska, M., Kornas, R., Cymerman, J., Panasiewicz, K., & Lipińska, H. (2020). Legume cover crops as one of the elements of strategic weed management and soil quality improvement. A Review. *Agriculture, 10*, 394. https://doi.org/10.3390/agriculture10090394

Kokkini, M., Gazoulis, I., Danaskos, M., Kontogeorgou, V., Kanatas, P., & Travlos, I. (2025). Enhancing ecosystem services in agriculture: The special role of legume intercropping. *Frontiers in Sustainable Food Systems, 9*, 1547879. https://doi.org/10.3389/fsufs.2025.1547879

Koriem, M., & Hamd-Alla, W. (2025). Impact of preceding crops and various fertilizer applications on sugar beet productivity, quality, and economic return. *Journal of Sohag Agriscience (JSAS), 10*(2), 182–200. https://doi.org/10.21608/jsasj.2025.464135

Kumar, S., Meena, R. M., Lal, R., et al. (2018). Role of legumes in soil carbon sequestration. In R. Meena, A. Das, G. Yadav, & R. Lal (Eds.), *Legumes for soil health and sustainable management*. Springer. https://doi.org/10.1007/978-981-13-0253-4_4

Lal, R. (2008). Carbon sequestration. *Philosophical Transactions of the Royal Society of London. Series B, Biological Sciences, 363*(1492), 815–830. https://doi.org/10.1098/rstb.2007.2185. PMID: 17761468; PMCID: PMC2610111.

Lal, R. (2017). Improving soil health and human protein nutrition by pulses-based cropping systems. *Advances in Agronomy, 145*, 167–204. https://doi.org/10.1016/bs.agron.2017.05.003

Lemage, B., & Tsegaye, M. (2020). Evaluation of legume shrubs improved fallow for abandoned agricultural land rehabilitation. *International Journal of Agricultural Research, Innovation and Technology, 10*, 64–70. https://doi.org/10.3329/ijarit.v10i1.48095

Lengwati, D. M., Mathews, C., & Dakora, F. D. (2020). Rotation benefits from N_2-fixing grain legumes to cereals: From increases in seed yield and quality to greater household cash-income by a following maize crop. *Frontiers in Sustainable Food Systems, 4*, 94. https://doi.org/10.3389/fsufs.2020.00094

Li, L., Sun, J. H., & Zhang, F. S. (2011). Intercropping with wheat leads to greater root weight density and larger below-ground space of irrigated maize at late growth stages. *Soil Science and Plant Nutrition, 57*, 61–67.

Liu, W. X., Wei, Y. X., Li, R. C., Chen, Z., Wang, H. D., Virk, A. L., et al. (2022). Improving soil aggregates stability and soil organic carbon sequestration by no-till and legume-based crop rotations in the North China Plain. *Science of the Total Environment, 847*, 157518. https://doi.org/10.1016/j.scitotenv.2022.157518

Liu, C., Feng, X., Xu, Y., et al. (2023). Legume-based rotation enhances subsequent wheat yield and maintains soil carbon storage. *Agronomy for Sustainable Development, 43*, 64. https://doi.org/10.1007/s13593-023-00918-4

Lüscher, A., Mueller-Harvey, I., Soussana, J. F., Rees, R. M., & Peyraud, J. L. (2014). Potential of legume-based grassland–livestock systems in Europe: A review. *Grass and Forage Science, 69*, 206–228. https://doi.org/10.1111/gfs.12124

Magnolo, F., Dekker, H., Decorte, M., Bezzi, G., Rossi, L., Meers, E., & Speelman, S. (2021). The role of sequential cropping and Biogasdoneright™ in enhancing the sustainability of agricultural systems in Europe. *Agronomy, 11*, 2102. https://doi.org/10.3390/agronomy11112102

Mattila, T. J., & Vihanto, N. (2024). Agricultural limitations to soil carbon sequestration: Plant growth, microbial activity, and carbon stabilization. *Agriculture, Ecosystems & Environment, 367*, 108986. https://doi.org/10.1016/j.agee.2024.108986

Mukhopadhyay, R., Sarkar, B., Jat, H. S., Sharma, B. C., & Bolan, N. S. (2021). Soil salinity under climate change: Challenges for sustainable agriculture and food security. *Journal of Environmental Management, 280*, 111736. https://doi.org/10.1016/j.jenvman.2020.111736

Nazir, M. J., Li, G., Nazir, M. M., Zulfiqar, F., Siddique, K. H. M., Iqbal, B., & Du, D. (2024). Harnessing soil carbon sequestration to address climate change challenges in agriculture. *Soil and Tillage Research, 237*, 105959. https://doi.org/10.1016/j.still.2023.105959

Oh, T. H. (2010). Carbon capture and storage potential in coal-fired plant in Malaysia—A review. *Renewable and Sustainable Energy Reviews, 14*(9), 2697–2709. https://doi.org/10.1016/j.rser.2010.06.003

Ouda, S., & Zohry, A. (2024a). Increasing land and water use efficiencies of wheat: Case study of Egypt. In *Integration of legume crops with cereal crops under changing climate: Sustainably increases food production*. Springer Publishing House. ISBN: 978-3-031-68101-1.

Ouda, S., & Zohry, A. (2024b). Legumes improve wheat and maize productivity grown in different cropping systems under changing climate. In *Integration of legume crops with cereal crops under changing climate: Sustainably increases food production*. Springer Publishing House. ISBN: 978-3-031-68101-1.

Panagea, I. S., Quataert, P., Alonso-Ayuso, M., Bárcena, T. G., De Boever, M., Diacono, M., Jacobs, A., Jensen, J. L., Seidel, F., Seitz, D., Spiegel, H., Vanden Nest, T., Don, A., & Ruysschaert, G. (2025). Forage vs. grain legumes: Contrasting effects on soil organic carbon stocks–evidence from 30 European field experiments. *European Journal of Soil Science, 76*(2), e70086. https://doi.org/10.1111/ejss.70086

Paustian, K., Lehmann, J., Ogle, S., Reay, D., Robertson, G. P., & Smith, P. (2016). Climate-smart soils. *Nature, 532*, 49–57. https://doi.org/10.1038/nature17174

Power, A. G. (2010). Ecosystem services and agriculture: Tradeoffs and synergies. *Philosophical transactions of the royal society B: biological sciences, 365*, 2959–2971.

Rani, K., Rani, A., Sharma, P., Dahiya, A., Punia, H., Kumar, S., Sheoran, S., & Banerjee, A. (2022). Legumes for agroecosystem services and sustainability. In *Advances in legumes for sustainable intensification* (pp. 363–380). Academic Press. https://doi.org/10.1016/B978-0-323-85797-0.00008-2

Restovich, S. B., Andriulo, A. E., & Portela, S. I. (2022). Cover crop mixtures increase ecosystem multifunctionality in summer crop rotations with low N fertilization. *Agronomy for Sustainable Development, 42*, 19. https://doi.org/10.1007/s13593-021-00750-8

Roohi, M., Arif, M. S., Guillaume, T., & Yasmeen, T. (2022). Role of fertilization regime on soil carbon sequestration and crop yield in a maize-cowpea intercropping system on low fertility soils. *Geoderma, 428*, 116152. https://doi.org/10.1016/j.geoderma.2022.116152

Stagnari, F., Maggio, A., Galieni, A., et al. (2017). Multiple benefits of legumes for agriculture sustainability: An overview. *Chemical and Biological Technologies in Agriculture, 4*, 2. https://doi.org/10.1186/s40538-016-0085-1

Stockmann, U., Adams, M. A., Crawford, J. W., Field, D. J., Henakaarchchi, N., Jenkins, M., Minasny, B., McBratney, A. B., De Remy de Courcelles, V., Singh, K., et al. (2013). The knowns, known unknowns and unknowns of sequestration of soil organic carbon. *Agriculture, Ecosystems & Environment, 164*, 80–99.

Tang, X., Zhang, C., Yu, Y., Shen, J., van der Werf, W., & Zhang, F. (2021). Intercropping legumes and cereals increase phosphorus use efficiency; a meta-analysis. *Plant and Soil, 460*, 89–104. https://doi.org/10.1007/s11104-020-04768-x

Thapa, R., Hanson, S., Hua, J., & Moore, V. M. (2025). Breeding for cold tolerance in common annual legume cover crops. *Crop Science*. https://doi.org/10.1002/csc2.70059

Tian, J. H., Kuang, X. Z., Tang, M. T., Chen, X. D., Huang, F., Cai, Y. X., & Cai, K. Z. (2021). Biochar application under low phosphorus input promotes soil organic phosphorus mineralization by shifting bacterial phoD gene community composition. *Science of the Total Environment, 779*, 146556.

Verhagen, W., van der Zanden, E. H., Strauch, M., van Teeffelen, A. J. A., & Verburg, P. H. (2018). Optimizing the allocation of agri-environment measures to navigate the trade-offs between ecosystem services, biodiversity and agricultural production. *Environmental Science & Policy, 84*, 186–196. https://doi.org/10.1016/j.envsci.2018.03.013

Wang, W., Li, M.-Y., Wang, Y., Li, J.-M., Zhang, W., et al. (2025). Legume intercropping improves soil organic carbon stability in drylands: A 7-year experimental validation. *Agriculture, Ecosystems & Environment, 381*, 109456. https://doi.org/10.1016/j.agee.2024.109456

Xue, L., Ren, H., Brodribb, T. J., Wang, J., Yao, X., & Li, S. (2020). Long term effects of management practice intensification on soil microbial community structure and co-occurrence network in a non-timber plantation. *Forest Ecology and Management, 459*, 117805. https://doi.org/10.1016/j.foreco.2019.117805

Zhang, D. B., Yao, Z. Y., Chen, J., et al. (2019). Improving soil aggregation, aggregate-associated C and N, and enzyme activities by green manure crops in the loess plateau of China. *European Journal of Soil Science, 70*, 1267–1279. https://doi.org/10.1111/ejss.12843

Zhang, K., Liu, Z., McCarl, B. A., & Fei, C. J. (2024). Enhancing agricultural soil carbon sequestration: A review with some research needs. *Climate, 12*, 151. https://doi.org/10.3390/cli12100151

Zhang, J., Hu, B., Zheng, Zhang, Z., Wang, B., Dong, K., & Dong, Y. (2025). Long term intercropping promotes improvement of soil quality and alleviates faba bean wilt disease. *Agriculture, Ecosystems & Environment, 381*, 109443. https://doi.org/10.1016/j.agee.2024.109443

Zhen, Y., Zhang, X., Zhang, C., Gao, Q., Dong, J., Zhang, L., Lu, X., & Wang, Y. (2023). Effects of climate change and land use/cover changes on carbon sequestration in forest ecosystems in the coastal area of China. *Frontiers in Forests and Global Change, 6*, 1271239. https://doi.org/10.3389/ffgc.2023.1271239

Zohry, A., & Ouda, S. (2025). Legume crops enhance water use efficiency under intercropping system with wheat. *Moroccan Journal of Agricultural Sciences, 6*(3), 145–160.

Zohry, A., Ouda, S., & Abdel-Wahab, T. (2020). Sustainable intensive cropping to reduce irrigation-induced erosion: I. Intercropping systems under surface irrigation practice. *Moroccan Journal of Agricultural Science, 1*(2), 63–71.

Zomer, R. J., Bossio, D. A., Sommer, R., & Verchot, L. V. (2017). Global sequestration potential of increased organic carbon in cropland soils. *Scientific Reports, 7*, 15554.

Zuber, S. M., Behnke, G. D., Nafziger, E. D., & Villamil, M. P. (2018). Carbon and nitrogen content of soil organic matter and microbial biomass under long-term crop rotation and tillage in Illinois, USA. *Agriculture, 8*, 37. https://doi.org/10.3390/agriculture8030037

Chapter 3
Decomposition of Legume Crop Residues: A Sustainable Pathway to Soil Conservation

Abstract The breakdown of legume crop residues is essential for sustaining soil health and improving ecosystem services in agroecosystems. This chapter examines the ecological and agronomic importance of legume residue decomposition in sustaining and enhancing soil health in agricultural systems. It examines the impact of legumes on the activity of soil microbial and invertebrates, nutrient cycling, and organic matter dynamics. Legumes significantly boost soil fertility and conservation, as well as ecosystem resilience, due to their high nitrogen content, biological nitrogen fixation, and interactions with soil microbes. This chapter synthesizes recent research to elucidate how legume residue management might serve as a sustainable approach to enhance soil productivity and promote climate-smart agriculture. Moreover, the significance of legumes in soil conservation against water and wind erosion was also discussed.

Keywords Leguminous residues breakdown · Soil microbial activity · Soil invertebrates · Soil conservation

Introduction

Agricultural output is significantly reliant on ecosystem services, including soil structure and fertility, nitrogen cycling, pollination, and pest management (Power, 2010). Agriculture and ecosystem services are interconnected through a mutually beneficial relationship. Agroecosystems provide essential services to society, including food production and soil conservation, while also depending on services supplied by natural ecosystems—such as pollination—that regulate and sustain the entire crop growth process (Palm et al., 2014). Enhancing these interconnected ecosystem services may augment food supply and diminish agriculture's reliance on external inputs, so effectively mitigating the conflict between agricultural productivity and environmental issues (Lazali & Drevon, 2023).

Lehmann and Kleber (2015) define residue decomposition as the process of disintegrating organic plant matter into smaller components, facilitated by abiotic

A. E. Zohry, S. A. H. Ouda, *The Agroecosystem Services of Leguminous Crops*, https://doi.org/10.1007/978-3-032-32461-0_3

factors (e.g., photodegradation) or by decomposer organisms via catabolism through microbial enzymes and/or invertebrate endogenous enzymes in conjunction with microbial symbionts. Soil microorganisms, namely bacteria and fungi, serve as the principal decomposers of agricultural residue (Hessen et al., 2004), whereas invertebrates such as millipedes and earthworms fragment and incorporate the residue into the soil, enhancing its accessibility for microorganisms (García-Palacios et al., 2013). The incorporation of crop residues into soils enhances microbial diversity and facilitates nutrient recycling, provides nitrogen, phosphorus, potassium, and carbon for succeeding crops, hence aiding in the reduction of greenhouse gas emissions (Diacono & Montemurro, 2010). Additionally, it enhances soil microbial activity, aeration, moisture retention, cation exchange capacity, and long-term carbon sequestration in soils (Choudhary et al., 2020). The quantity of organic matter in soil is influenced by the quality and quantity of carbon contributed and digested by soil microbes, as well as soil-forming elements such as parent material, climate, time, terrain, and organisms (Sissoko & Kpomblekou, 2010).

Conversely, invertebrate decomposers can influence plants through trophic and/or non-trophic mechanisms. The trophic effect impacts soil nutrient mineralization due to enzymatic breakdown in the gut, specifically affecting the nutritional composition of the decomposing substrate and/or soil (Winck et al., 2020). The non-trophic consequences of movement and nest building modify soil structure, namely soil particle distribution, aeration, and moisture levels, which can subsequently affect soil chemical composition (Tuma et al., 2019). Invertebrate decomposers physically disintegrate dead plants into smaller fragments, so enhancing the surface area for fungi and bacteria, while their digestive processes and excrement directly release nutrients (Harmon et al., 2020).

The breakdown rate of crop residues is primarily affected by soil texture and the C/N ratio of the residues. In soils with a dense texture, clays often shield organic particles from microbial degradation, hence diminishing decomposition rates as organic carbon is adsorbed onto soil minerals (Khalil et al., 2005). In sandy soils, bacteria have unrestricted access to organic particles, which they degrade swiftly, hence accelerating decomposition rates. Sandy soils demonstrate a limited ability to connect with organic molecules, whereas clay particles that interact with organic molecules are adsorbed by robust ligand exchange at several reactive sites (Cai et al., 2016).

The decomposition of legume crop residues in agroecosystems offers essential ecosystem services by recycling soil nutrients, augmenting soil organic matter and structure, and promoting soil fertility via the release of atmospheric nitrogen (Lazali & Drevon, 2023). This biological process decomposes complex organic materials into simpler molecules, enhancing the availability of vital nutrients such as nitrogen, phosphate, and potassium for succeeding crops and fostering microbial activity (Andreotti et al., 2018). High-quality legume residues, owing to their nitrogen concentration, enhance the soil C/N ratio, hence promoting overall soil health and fertility (Jensen et al., 2012). Deep-rooted legumes promote nitrogen solubilization through root exudates, hence enhancing nutrient uptake and water infiltration into the soil (Lazali & Drevon, 2023). Moreover, legumes augment invertebrate role in the decomposition process chiefly by supplying nitrogen-rich organic matter that invigorates the activity and biomass of decomposer organisms. The heightened invertebrate activity subsequently expedites the total decomposition and nutrient cycling within the soil ecosystem (Milcu et al., 2008a). Furthermore, some legumes have the ability to increase microbial population and activity in the soils, such as alfalfa, in which their plants have the ability to release superior-quality substrates from the roots compared to grasses (Dhakal & Islam, 2018).

This chapter analyzed the essential function of legumes in promoting agricultural sustainability via their contributions to ecosystem services, breakdown of plant wastes, and soil conservation. It specifically seeks to investigate the impact of legume-based cropping systems on soil microbial and invertebrate activity, atmospheric nitrogen fixation and cycling, nutrient availability, organic matter buildup, soil structure stabilization, and mitigation of soil erosion, thus fostering sustainable agricultural productivity and environmental resilience.

Legumes Augment Soil Microbial Activities

Microbial activity is a critical and sensitive indicator of soil health, as microbes facilitate nutrient cycling (Ferreira et al., 2013). The incorporation of legumes into cropping systems enhances microbial activity in the soil due to the superior substrate quality of legume rhizodeposits, characterized by soil moisture (Nair & Ngouajio, 2012) and C/N ratio of the residue, where lower C/N ratios promote faster degradation (Kebede, 2021). According to Lazali and Drevon (2023), legume residues enhance soil physical, chemical, and biological qualities, as well as facilitate the decomposition of residue due to their elevated nitrogen content, which accelerates microbial decomposition rates relative to other plant species (Ferreira et al., 2013).

According to Yuvaraj et al. (2020), legumes enhance the diversity of soil flora and fauna, hence augmenting the overall stability of soil life. Legume-based intercropping systems can enhance rhizobacterial community variety and soil health by promoting both symbiotic and non-symbiotic beneficial populations. Additionally, legumes promote a varied and active soil microbiome, crucial for nutrient cycling and overall soil health, hence significantly affecting the resilience of cultivated crops (Yuvaraj et al., 2020). Research indicates that cropping systems using legumes enhance the prevalence of microorganisms involved in soil nitrogen cycling, hence substantially modifying the content and organization of microbial communities (Luo et al., 2022). The alteration in microbial biomass and community composition would affect microbial enzyme activities, which are frequently utilized as indicators of microbial functions associated with organic matter decomposition and the enhancement of soil nutrient cycling (Meng et al., 2024).

Legume residues can provide a greater source of nitrogen for subsequent crops than cereal residues, owing to their higher nitrogen content and lower C/N ratio (Matse et al., 2020). The lower C/N ratio of legumes, when combined with non-legume residues, they may enhance the degradation process by promoting increased microbial diversity and nitrogen availability (Kohmann et al., 2019). The improvement of soil fertility via the decomposition of legume residues after harvest promotes the release of nitrogen and other nutrients for subsequent crops, hence enhancing their resilience in the face of climate change (Masson-Boivin & Sachs, 2018).

Legumes in intercropping systems augment soil microbial activity by elevating the biomass, diversity, and functionality of beneficial soil organisms, such as nitrogen-fixing bacteria and arbuscular mycorrhizal fungus (Ghani et al., 2003). Legume intercropping presents an approach to diminish nitrogen immobilization by concurrently integrating crop components with elevated nitrogen levels (Naresh et al., 2017). Upon reintroduction to agricultural land, the fresh legume residues facilitate nucleation essential for the development of macro-aggregates, which serve as a conducive environment for soil microorganisms (Liang, 2013). The bacterial community in the rhizosphere is essential for enhancing the growth and health of intercrops through many direct and indirect mechanisms involving plant growth-promoting rhizobacteria (Chamkhi et al., 2022). Moreover, the decomposition of organic residues in intercropping systems promotes the root exudation of enzymes such as phosphatases, phytase, and carboxylates (Rojas-Downing et al., 2017), which subsequently facilitates the mobilization and absorption of vital nutrients including phosphorus, iron, manganese, and zinc (Xiao et al., 2013).

According to Tang et al. (2021), microbial biomass carbon was found to be elevated in the rhizosphere just for intercropped durum wheat with chickpea and durum wheat with lentil. An increase in phosphorus availability in the rhizosphere was seen in response to intercropping in one of three crop species. According to Osunde et al. (2004) soybeans provide a favorable residual nitrogen effect on soil in subsequent seasons, attributed to their capacity to fix atmospheric nitrogen via symbiotic associations with root nodule bacteria. Zhang et al. (2024) assert that soybeans enhance soil health by augmenting organic matter and potentially enhancing soil structure, hence indirectly benefiting succeeding crops.

The amounts of nitrogen fixed and returned to the soil as crop residue differ from one legume crop to another. Faba beans and lentils may fix up to 140 kg N/ha of atmospheric nitrogen, whereas chickpeas can fix up to 160 kg N/ha, according to Kirihetti (2018). The total amounts of fixed nitrogen that were returned to the soil as crop residue were determined to be highest after cultivating chickpeas, while the lowest values were observed after cultivating faba beans and lentils, as stated by Kirihetti (2018). Additionally, it has been proposed that the subsequent crop benefits from the growth-promoting chemicals released by decaying legume crop residues (Arcand et al., 2014). Furthermore, according to Amossé et al. (2014), intercropping legumes with sugar beets reduces competition for nitrogen acquisition between species, enriches soil nitrogen, and produces a high-nitrogen source from leftovers. Additionally, it has been noted that chickpea plants yield a significant amount of residual nitrogen, which can be used as organic manure to increase soil fertility (Banik et al., 2018). Thus, legumes enhance the soil food web, facilitating the development of stable soil organic carbon and improving soil structure,

Table 3.1 Percentage improvement in the yield derived from various reported intercropping systems experiments in Egypt

Intercropping system	Increase in yield (%)	References
50% cowpea+100% maize	3%	Taha and Kasem (2022)
15% fahl clover*+100% wheat	8%	Ouda and Zohry (2024)
25% fahl clover*+100% wheat	6%	Zohry and Ouda (2025)
25% faba bean+100% sugar beet	2%	Ouda and Zohry (2023)
25% faba bean+100% wheat	1%	Zohry and Ouda (2025)
33% peas+100% wheat	2%	Zohry and Ouda (2025)
25% chickpea+100% sugar beet	2%	Ouda and Zohry (2023)
25% lentil+100% sugar beet	2%	Ouda and Zohry (2023)

*A variety of Egyptian clover

water retention, and general fertility, so resulting in more resilient and productive agricultural systems.

Nyawade et al. (2019) conducted a two-year field experiment to evaluate the effects of intercropping potato with two legumes, lima bean and dolichos. At the end of each growing season, crop residues from the different systems were measured and incorporated into the soil at the start of the subsequent season. The results showed that, compared with sole potato cultivation, intercropping significantly increased dissolved organic matter by 7–21% and microbial biomass by 15–38%, which in turn stimulated soil enzyme activities. Patterns of soil microbial respiration closely paralleled enzyme activity, with intercropped systems exhibiting 20–34% higher respiration rates than monocropped potato. These findings suggest that intercropping enhances the return of organic residues, which are temporarily retained within soil macro-aggregates, thereby ensuring a more continuous supply of substrates to support soil microbial activity.

Furthermore, in other trials conducted by Wang et al. (2025), potato-soybean intercropping and monoculture trials over seven consecutive years to examine the formation and transformation of particulate organic carbon, mineral-associated organic carbon, microbial necromass, and soil organic carbon stability in relation to microbial and enzyme stoichiometry. Their findings indicated that, in contrast to potato monoculture, intercropping markedly reduced free particulate organic carbon concentration by 13% and enhanced mineral-associated organic carbon content by 18%. Furthermore, intercropping enhanced soil organic carbon content by 8% and augmented its stability by 23%.

The increase in the yield of the intercrops of legume intercropping systems in the following season is a strong indication of the decomposition and mineralization of the legume residues, which release beneficial nitrogen and other nutrients into the soil. In Egypt, implemented intercropping systems with legumes resulted in an increase in the yield of the companion crop in the following growing season, compared to the first growing season. Table 3.1 presented the percentage of increase in the yield in the second growing season derived from various reported intercropping systems experiments in Egypt.

Legumes Augment the Actions of Soil Invertebrates

Soil invertebrates are essential for maintaining soil physical qualities, facilitating short-term nutrient recycling, and ensuring long-term preservation of nutrients and organic matter inside soil structures (Sileshi & Mafongoya, 2006). Soil invertebrates are earthworms, collembola, mites, enchytraeids, and other macro- and mesofauna (Sileshi & Mafongoya, 2006). Soil invertebrates operate as a community that facilitates essential soil activities, including organic matter breakdown and nitrogen cycling (Bardgett & Wardle, 2010). Augmenting plant diversity can improve ecosystem functionality (including plant productivity, organic matter decomposition, and nutrient cycling) by optimizing resource utilization or fostering advantageous interspecies interactions (Barry et al., 2019). Increased soil faunal variety improves the stability of decomposition and nutrient cycling processes via functional redundancy, especially under environmental stress (Thakur et al., 2015).

The incorporation of legumes into an ecosystem typically results in increased invertebrate variety, abundance, and functionality, especially within agricultural systems (Perera et al., 2022). A primary mechanism driving these advantages is the enhancement in both the amount and quality of organic inputs linked to legumes. In comparison to cereals, legume wastes typically exhibit elevated nitrogen contents and diminished C/N ratios, rendering them more amenable to decomposition and appealing to soil fauna (Schmidt, 2002).

Legumes affect soil invertebrates via their live roots and rhizosphere interactions. Biological nitrogen fixation brings nitrogen into agroecosystems, generating nutrient-rich microsites that enhance microbial biomass and activity (Wardle et al., 2004). Furthermore, legume roots exude labile carbon molecules that stimulate microbial proliferation and sustain elevated trophic levels within the soil food web (Milcu et al., 2008a). Legume intercropping systems generally sustain more diversity and continuity of root inputs throughout the growing season, enabling soil invertebrate populations to thrive and reproduce more efficiently than in simplified monocultures (Perera et al., 2022). Extended field investigations demonstrate that legume-based intercropping and cover cropping systems regularly foster more diverse and balanced soil invertebrate communities compared to monoculture systems (Blanchart et al., 2006).

Functional interactions between legumes and soil invertebrate decomposers frequently produce synergistic feedbacks that improve agroecosystem performance. Earthworms function as ecosystem engineers by intermixing organic and mineral soil components, enhancing aggregation, and augmenting nutrient accessibility for crops (Lavelle et al., 2016). Research in temperate and tropical ecosystems indicates that earthworm abundance and biomass are consistently greater in legume-inclusive cropping systems compared to continuous monocropping (Blanchart et al., 2006). Augmented earthworm activity expedites trash fragmentation and its integration into the soil, thus elevating decomposition rates and nutrient cycling (Milcu et al., 2008a). High-quality residues and root inputs from legumes enhance earthworm-mediated processes, resulting in increased nitrogen mineralization and enhanced soil physical characteristics. Experimental investigations indicate that the synergistic presence of legumes and earthworms can augment crop growth beyond the impact of nutrient inputs alone, underscoring the significance of biological interactions in legume intercropping systems (Milcu et al., 2008b).

When intercropped, legumes became an agricultural method that utilizes the distinct ecosystem services of legumes to improve the resilience and productivity of cropping systems (Kokkini et al., 2025). Legumes in intercropping systems frequently enhance both the quantity and quality of organic inputs to soil (increased litter and root biomass, augmented nitrogen-rich residues), which consistently benefits soil invertebrate decomposers. These organisms are crucial facilitators of organic matter breakdown, nutrient mineralization, and soil structure development (Wardle et al., 2004). Research across several agroecosystems demonstrates that the incorporation of legumes into intercropping systems increases the diversity and functional efficacy of soil invertebrate communities (Lavelle et al., 2016). Liu et al. (2025c) reported that legume intercropping generally has a positive and complex effect on invertebrate decomposers by enhancing the soil environment and increasing organic matter inputs, though specific effects vary by the type of decomposer and the cropping system. Various legume species exhibit differences in root architecture, phenology, and residue chemistry, resulting in a mosaic of niches that sustains various soil invertebrate assemblages (Blanchart et al., 2006). In cereal-legume intercrops, residue mixes generate a more equilibrated substrate that promotes microbial proliferation and offers superior nutritional resources for macrofauna, including earthworms (Milcu et al., 2008b).

Improvements in microclimate and soil characteristics linked to legume intercropping enhance beneficial faunal responses (Akchaya et al., 2025). Enhanced ground cover and residue retention in intercropped systems mitigate soil temperature variations and improve moisture retention, characteristics essential for the survival and activity of several soil invertebrates (Blanchart et al., 2006). Earthworms are particularly vulnerable to desiccation and thrive in the enhanced soil moisture and aggregation typically found in legume-based systems (Milcu et al., 2008b). Mesofauna, including collembola, have positive responses to enhanced pore connectivity and improved organic matter availability resulting from legume-induced soil structural enhancements (Thakur et al., 2015).

From the standpoint of biodiversity and resilience, legume intercropping systems enhance habitat variety and resource diversity in the soil. Thus, integrating legumes into intercropping systems constitutes a substantiated approach for augmenting belowground biodiversity and the associated ecosystem services.

Essential Ecosystem Services Offered by Legume Residues

Improve Soil Composition and Nutrient Content

The decomposition of residue releases organic molecules that adhere soil particles, resulting in the formation of stable aggregates. Legumes enhance soil structure by contributing organic matter, disaggregating soil aggregates, and increasing porosity via their root systems, root exudates, and residues (Smýkal et al., 2014). Root exudates, including sugars, amino acids, and organic acids, provide nourishment for soil microbes, thereby enhancing microbial activity and facilitating the development of stable soil aggregates that safeguard organic matter from decomposition, augment carbon sequestration, and contribute to soil fertility and resilience (Kebede, 2021). Legume-based cropping systems can enhance macro- and micro-aggregates for soil structure by 52–111% compared to sole cropping (Chamkhi et al., 2022). Moreover, the integration of root biomass into the soil increases organic carbon levels, so augmenting soil fertility and enhancing its capacity to store water and nutrients, which renders it more tolerant to climate change effects (Duarte et al., 2024).

The improvement in soil fertility through a legume-based cropping strategy positively influences the production of cultivated crops. Ouda et al. (2020) observed an 11% improvement in wheat yield when preceded by an early winter crop, fahl clover, compared to the yield obtained when preceded by maize. Furthermore, wheat yield was increased by 28% when preceded by cowpea intercropped with maize, followed by fahl clover as an early winter crop, compared to the yield obtained when preceded by just by maize. Zohry et al. (2020) intercropped peas with wheat in the winter season and cowpea intercropped with sunflower in the summer season. Their findings indicated a 5% improvement in wheat output compared to sole wheat cultivation and a 12% increase in sunflower yield, compared to the yield from sole sunflower cultivation.

Atmospheric Nitrogen Fixation

Legumes improve soil health by fixing atmospheric nitrogen, augmenting nutrient availability, and promoting beneficial microbial activity in the soil (Nazir et al., 2024). A portion of the symbiotically fixed nitrogen in legumes becomes accessible to subsequent crops via the decomposition and mineralization of legume residues (Islam & Adjesiwor, 2017). Furthermore, the augmentation of biological nitrogen fixation using rhizobia inoculants enhances the availability of nitrogen and phosphorus in the soil post-harvest, which can be utilized by subsequent crops (Matse et al., 2020). Bhowmik and Das (2018) assert that biological nitrogen fixation can be enhanced by inoculating host legumes with compatible rhizobia, resulting in residual nitrogen in the soil, contributing to organic matter, and serving as a cost-effective nutrient source for subsequent cropping seasons.

Intercropping with legumes has been shown to improve soil nitrogen via plant residues (Amossé et al., 2014). Furthermore, the concentrations of soil accessible nitrogen were elevated in the second season of legume intercropping system, compared to the first season, attributable to the decomposition of leguminous crop residues that enhance the succeeding crop (Arcand et al., 2014).

Ghosh et al. (2007) determined that legumes, including soybean, pigeon pea, cowpea, and groundnut, when intercropped with maize, positively influenced the yield of the subsequent wheat crop. It has been reported that an elevated level of soil accessible nitrogen was observed in the faba bean intercropping system with wheat, which is attributed to its superior nitrogen fixation compared to peas and chickpea (Carranca et al., 1999), as well as its robust root systems (Shanmugam et al., 2022). Bargaza et al. (2015) suggested that intercropping with faba bean allows its roots to penetrate deeper soil layers, consequently accessing a larger soil volume (Belachew et al., 2019), perhaps enhancing organic matter decomposition in the soil (Arcand et al., 2014). Carlsson et al. (2017) indicated that in a pea–wheat intercropping system, wheat exhibits greater competitiveness for mineral nitrogen, compelling peas to enhance atmospheric nitrogen fixation through symbiosis to satisfy its requirements, hence impeding weed development owing to nitrogen scarcity.

In Egypt, Ouda and Zohry (2023) executed a two-year field experiment to investigate the impact of intercropping faba bean and chickpea with sugar beet on available soil nitrogen. They concluded that the faba bean intercropping system resulted in a 12% increase in available soil nitrogen, while the chickpea intercropping system achieved a 13% increase, relative to the levels observed under sole sugar beet cultivation. In another research by Zohry and Ouda (2025), faba bean, peas, and fahl clover were intercropped with wheat, resulting in increases in accessible soil nitrogen of 21%, 15%, and 3%, respectively, compared to the value obtained from wheat monoculture.

Nutrient Availability and Uptake

The availability and uptake of plant nutrients mostly depend on the quantity, concentrations, and activities occurring in the soil's root zone, as well as the soil's capacity to replenish limiting nutrients in the soil solution (Makoi et al., 2013). Nitrogen and phosphorus are the primary elements abundantly present in soil, yet exist in forms that are inaccessible to plants. These essential elements typically serve as the limiting variables for plant growth and development (Mmbaga et al., 2014). Legume-based cropping systems enhance various facets of soil fertility, including soil organic carbon and humus levels, nitrogen and phosphorus availability and absorption, and organic carbon and nitrogen, while also producing hydrogen gas as a by-product of biological nitrogen fixation, which fosters the development of bacterial nodules in the rhizosphere (Stagnari et al., 2017). The decomposition of legume residues naturally replenishes soil nutrients, hence decreasing dependence on synthetic fertilizers, enhancing agroecosystem sustainability, and lowering related expenses. Legumes can alter their rhizosphere while fixing nitrogen by releasing organic acids, such as psionic acid and citric acid, which enhance phosphorus availability for both the legumes and subsequent crops (Ahmadi & Shafiqi, 2023).

Soil pH is acknowledged as a crucial determinant of soil biochemical processes (Philippot et al., 2024). Legumes can alter rhizosphere pH, so producing a microenvironment that improves nutrient availability for neighboring intercrops (Goyal & Habtewold, 2023). Biological nitrogen fixation releases hydrogen gas as a by-product, facilitating bacterial proliferation in the rhizosphere (Li et al., 2020). Moreover, Ferguson et al. (2013) revealed that legumes may more efficiently remove calcium and magnesium from the soil than cereals, substituting these cations with hydrogen ions. This exchange results in the elimination of OH^- ions and a rise in H^+ concentration, ultimately decreasing soil pH and aiding in the mitigation of N_2O emissions (Hénault et al., 2019).

The Function of Legumes in Soil Conservation

Soil conservation involves practices to prevent the topmost layer of soil from eroding or losing fertility due to wind, water, or overuse, ensuring land remains productive for agriculture, supporting ecosystems (Rodríguez et al., 2022). Soil conservation is vital for climate change adaptation, maintaining soil health, and facilitating crop growth by supplying key minerals. Nevertheless, severe weather phenomena, including intense precipitation and high winds, might exacerbate soil erosion (Akchaya et al., 2025).

Intercropping has demonstrated efficacy in addressing these difficulties, particularly when integrated with conservation techniques such as cover cropping and mulching, which can diminish erosion by as much as 50% (Lal, 2017). Liu et al. (2025a, b, c) assert that intercropping with legumes diminishes runoff by establishing a more intricate and dense plant cover that intercepts precipitation and enhances soil structure, facilitating water infiltration into the ground rather than surface flow. Tamta et al. (2019) reported that intercropping cowpeas with maize (two rows of maize alongside one row of cowpea) diminished runoff by 10% and soil erosion by 28% in comparison to maize monoculture. Additionally, intercropping soybean, peanuts, or cowpea with maize, sorghum, or pearl millet has been efficient in mitigating soil erosion (Akchaya et al., 2025). Moreover, Fan et al. (2016) noted diminished runoff rates in intercropped maize and potato systems compared to their respective monoculture systems. Liu et al. (2025a, b) documented a 29% decrease in runoff associated with legume intercropping and a notable reduction in soil erosion.

Soil conservation can be accomplished through the use of cover crops to protect soil and reduce erosion (Huang et al., 2015). Cover crops are grown between main crops to improve soil health and protect the land from erosion during the off-season

(Akchaya et al., 2025). Cover crops can mitigate soil degradation caused by wind and water erosion. Water erosion impacts both soil and water quality, while wind erosion primarily influences soil quality by diminishing its fertility (De Baets et al., 2011). The efficacy of legume cover crops in mitigating soil water erosion is contingent upon the rate of plant growth and biomass production (Kocira et al., 2020).

Legume cover crops mitigate wind erosion, which negatively impacts soil quality, particularly in semi-arid regions during periods of strong winds when plant cover is lacking (late winter and early spring) (Kocira et al., 2020). The implementation of winter or spring cover crops that thrive in these conditions mitigates wind erosion. Furthermore, legumes function as beneficial cover crops, offering advantages like nitrogen fixation, weed suppression, and soil preservation (Gazoulis et al., 2025). Employing legumes as cover crops during fallow periods improves soil health, reduces erosion and nitrate leaching, facilitates water infiltration, and retains soil moisture (Kocira et al., 2020). Furthermore, combinations of legume cover crops with other species (e.g., grasses) are more efficacious in enhancing uniform surface cover compared to monoculture cultivation, hence significantly mitigating soil erosion (Wortman et al., 2012). Blanco-Canqui and Ruis (2020) noted a decrease in vulnerability to soil wind erosion by the cultivation of cover crops (such as legumes) during fallow periods in wheat crop rotation.

The cultivation of legume cover crops mitigates wind erosion risk by enhancing soil physical properties, improving structure, augmenting organic carbon content, and increasing soil aggregate retention through the root systems of cover crops (Kocira et al., 2020). According to De Baets et al. (2011), augmenting the organic carbon content in soil exerts an anti-erosion influence, enhances the stability of aggregates, and diminishes the proportion of soil vulnerable to wind erosion, as organic carbon may physically, chemically, and biologically bind soil particles to create stable macro-aggregates. Liu et al. (2025c) reported that maize intercropped with cover crops exhibited substantial decreases in runoff and soil erosion, particularly when the cover crops possessed elevated stem density.

Conclusion

Legumes are critical for enhancing soil health, fertility, and structure, thereby offering numerous ecosystem services vital for sustainable agriculture. Legumes enhance soil quality by fixing atmospheric nitrogen, so increasing organic matter and nutritional content, which diminishes reliance on synthetic fertilizers. Their remnants augment microbiological activity, nitrogen cycling, and carbon sequestration, while their extensive root systems enhance soil structure and stability, reducing erosion and facilitating water infiltration. Intercropping and cover cropping with legumes have demonstrated the ability to augment soil microbial diversity, promote nutrient availability, and elevate crop yields. Integrating legumes into agricultural systems presents a sustainable and environmentally beneficial approach to enhancing soil productivity, reducing environmental degradation, and fostering resilient agroecosystems amid climate change.

References

Ahmadi, A. Y., & Shafiqi, S. (2023). Relations of legumes with soil health and succeeding crops. *Journal of Agricultural Research, 8*(4), 000333.

Akchaya, K., Parasuraman, P., Pandian, K., Vijayakumar, S., Thirukumaran, K., Mustaffa, M. R. A. F., Rajpoot, S. K., & Choudhary, A. K. (2025). Boosting resource use efficiency, soil fertility, food security, ecosystem services, and climate resilience with legume intercropping: A review. *Frontiers in Sustainable Food Systems, 9*, 1527256. https://doi.org/10.3389/fsufs.2025.1527256

Amossé, C., Jeuffroy, M. H., Mary, B., & David, C. (2014). Contribution of relay intercropping with legume cover crops on nitrogen dynamics in organic grain systems. *Nutrient Cycling in Agroecosystems, 98*, 1–14. https://doi.org/10.1007/s10705-013-9591-8

Andreotti, F., Mao, D., Jagoret, P., Speelman, E. N., Gary, C., & Saj, S. (2018). Exploring management strategies to enhance the provision of ecosystem services in complex smallholder agroforestry systems. *Ecological Indicators, 94*, 257–265.

Arcand, M. M., Lemke, R., Farrell, R. E., & Knight, J. D. (2014). Nitrogen supply from belowground residues of lentil and wheat to a subsequent wheat crop. *Biology and Fertility of Soils, 50*(3), 507–515.

Banik, D., Upadhyay, H., Devi, B. P., & Darvhankar, M. (2018). Effect of different irrigation system on growth and development of Wheat-Chickpea intercropping. *Bulletin of Environmental Pharmacology and Life Sciences, 8*(1), 32–34.

Bardgett, R. D., & Wardle, D. A. (2010). *Aboveground-belowground linkages: Biotic interactions, ecosystem processes, and global change*. Oxford University Press.

Bargaza, A., Isaacb, M. E., Jensena, E. S., & Carlsson, G. (2015). Intercropping of faba bean with wheat under low water availability promotes faba bean nodulation and root growth in deeper soil layers. *Procedia Environmental Sciences, 29*, 111–112. https://doi.org/10.1016/j.proenv.2015.07.188

Barry, K. E., Mommer, L., van Ruijven, J., Wirth, C., Wright, A. J., Bai, Y., Connolly, J., De Deyn, G. B., de Kroon, H., Isbell, F., Milcu, A., Roscher, C., Scherer-Lorenzen, M., Schmid, B., & Weigelt, A. (2019). The future of complementarity: Disentangling causes from consequences. *Trends in Ecology and Evolution, 34*, 167–180. https://doi.org/10.1016/j.tree.2018.10.013

Belachew, K. Y., Nagel, K. A., Poorter, H., & Stoddard, F. L. (2019). Association of shoot and root responses to water deficit in young faba bean (*Vicia faba L.*) plants. *Frontier in Plant Science, 10*, 1063. https://doi.org/10.3389/fpls.2019.01063

Bhowmik, N. S., & Das, A. (2018). Biofertilizers: A sustainable approach for pulse production. In R. Meena, A. Das, G. Yadav, & R. Lal (Eds.), *Legumes for soil health and sustainable management* (pp. 445–485). Springer. https://doi.org/10.1007/978-981-13-0253-4_14

Blanchart, E., Villenave, C., Viallatoux, A., Barthès, B., Girardin, C., Azontonde, A., & Feller, C. (2006). Long-term effect of a legume cover crop on soil macrofauna in a tropical humid environment. *Soil Biology & Biochemistry, 38*(8), 216–227.

Blanco-Canqui, H., & Ruis, S. J. (2020). Cover crop impacts on soil physical properties: A review. *Soil Science Society of America Journal, 84*, 1527–1576. https://doi.org/10.1002/saj2.20129

Cai, A., Xu, H., Shao, X., Zhu, P., Zhang, W., Xu, M., et al. (2016). Carbon and nitrogen mineralization in relation to soil particle-size fractions after 32 years of chemical and manure application in a continuous maize cropping system. *PLoS One, 11*(3), e0152521. https://doi.org/10.1371/journal.pone.0152521

Carlsson, G., Bedoussac, L., Cupina, B., Djordjevic, V., Gaudio, N., Jensen, E.-S., Jeuffroy, M.-H., Journet, E.-P., Justes, E., & Mikic, A. (2017). does a mixture of pea varieties with different leaf morphology improve crop performance? In *Proceedings of the international conference on advances in grain legume cultivation and use, Novi Sad, Serbia, 27–28 September 2017*.

Carranca, C., de Varennes, A., & Rolston, D. (1999). Biological nitrogen fixation by faba bean, pea and chickpea, under field conditions, estimated by the 15^N isotope dilution technique. *European Journal of Agronomy, 10*(1), 49–56. https://doi.org/10.1016/S1161-0301(98)00049-5

Chamkhi, I., Cheto, S., Geistlinger, G., Zeroual, Y., & Kouisni, L. (2022). Legume-based intercropping systems promote beneficial rhizobacterial community and crop yield under stressing conditions. *Industrial Crops and Products, 183*, 114958. https://doi.org/10.1016/j.indcrop.2022.114958

Choudhary, V. K., Gurjar, D. S., & Meena, R. S. (2020). Crop residues and weed biomass incorporation with microbial inoculation improve the crop and soil productivity in the rice (Oryza Sativa L.) -toria (Brassica rapa L.) cropping system. *Ecology*, article 100048. https://doi.org/10.1016/j.indic.2020.100048

De Baets, S., Poesen, J., Meersmans, J., & Serlet, L. (2011). Cover crops and their erosion-reducing effects during concentrated flow erosion. *Catena, 85*, 237–244.

Dhakal, D., & Islam, M. A. (2018). Grass-legume mixtures for improved soil health in cultivated agroecosystem. *Sustainability, 10*, 2718–2729.

Diacono, M., & Montemurro, F. (2010). Long-term effects of organic amendments on soil fertility. *Agronomy for Sustainable Development, 30*, 401–422. https://doi.org/10.1007/978-94-007-0394-0_34

Duarte, R. D. C., Iannetta, P. P. M., Gomes, A. M., & Vasconcelos, M. W. (2024). More than a meat or synthetic nitrogen fertiliser-substitute: A review of legume phytochemicals as drivers of 'one health' via their influence on the functional diversity of soil- and gut-microbes. *Frontiers in Plant Science, 15*, 1337653. https://doi.org/10.3389/fpls.2024.1337653

Fan, Z., An, T., Wu, K., Zhou, F., Zi, S., Yang, Y., et al. (2016). Effects of intercropping of maize and potato on sloping land on the water balance and surface runoff. *Agricultural Water Management, 166*, 9–16. https://doi.org/10.1016/j.agwat.2015.12.006

Ferguson, B. J., Lin, M. H., & Gresshoff, P. M. (2013). Regulation of legume nodulation by acidic growth conditions. *Plant Signaling & Behavior, 8*(3), e23426.

Ferreira, A. S., Dos Santos, M. A., & Corrêa, G. F. (2013). Soil microbial response to glucose and phosphorus addition under agricultural systems in the Brazilian Cerrado. *Anais da Academia Brasileira de Ciências, 85*(1), 395–403. https://doi.org/10.1590/s0001-37652013005000021

García-Palacios, P., Maestre, F. T., Kattge, J., & Wall, D. H. (2013). Climate and litter quality differently modulate the effects of soil fauna on litter decomposition across biomes. *Ecology Letters, 16*, 1045–1053.

Gazoulis, K. M. I., Kontogeorgou, D. M. V., Kanatas, P., & Travlos, I. (2025). Enhancing ecosystem services in agriculture: The special role of legume intercropping. *Frontiers in Sustainable Food Systems, 9*, 1547879. https://doi.org/10.3389/fsufs.2025.1547879

Ghani, A., Dexter, M., & Perrott, K. W. (2003). Hot-water extractable carbon in soils: A sensitive measurement for determining impacts of fertilization, grazing and cultivation. *Soil Biology & Biochemistry, 35*(9), 1231–1243.

Ghosh, P. K., Bandyopadhyay, K. K., Wanjari, R. H., Manna, M. C., Misra, A. K., Mohanty, M., et al. (2007). Legume effect for enhancing productivity and nutrient use-efficiency in major cropping systems–an Indian perspective: A review. *Journal of Sustainable Agriculture, 30*, 59–86. https://doi.org/10.1300/J064v30n01_07

Goyal, R. K., & Habtewold, J. Z. (2023). Evaluation of legume-rhizobial symbiotic interactions beyond nitrogen fixation that help the host survival and diversification in hostile environments. *Microorganisms, 11*(6), 1454. https://doi.org/10.3390/microorganisms11061454

Harmon, M. E., Fasth, B. G., Yatskov, M., Kastendick, D., Rock, J., & Woodall, C. W. (2020). Release of coarse woody detritus-related carbon: A synthesis across forest biomes. *Carbon Balance and Management, 15*, 1–21.

Hénault, C., Bourennane, H., Ayzac, A., et al. (2019). Management of soil pH promotes nitrous oxide reduction and thus mitigates soil emissions of this greenhouse gas. *Scientific Reports, 9*, 20182. https://doi.org/10.1038/s41598-019-56694-3

Hessen, D. O., Ågren, G. I., Anderson, T. R., Elser, J. J., & Ruiter, P. C. (2004). Carbon sequestration in ecosystems: The role of stoichiometry. *The Ecologist, 85*, 1179–1192. https://doi.org/10.1890/02-0251

Huang, J., Tichit, M., Poulot, M., Darly, S., Li, S. C., Petit, C., & Aubry, C. (2015). Comparative review of multifunctionality and ecosystem services in sustainable agriculture. *Journal of Environmental Management, 149*, 138–147.

Islam, M. A., & Adjesiwor, A. T. (2017). Nitrogen fixation and transfer in agricultural production systems. In Amanullah & S. Fahad (Eds.), *Nitrogen in agriculture-updates*. IntechOpen. https://doi.org/10.5772/intechopen.71766

Jensen, E. S., Peoples, M. B., Boddey, R. M., et al. (2012). Legumes for mitigation of climate change and the provision of feedstock for biofuels and biorefineries. A review. *Agronomy for Sustainable Development, 32*, 329–364. https://doi.org/10.1007/s13593-011-0056-7

Kebede, E. (2021). Contribution, utilization, and improvement of legumes-driven biological nitrogen fixation in agricultural systems. *Frontiers in Sustainable Food Systems, 5*, 1–18. https://doi.org/10.3389/fsufs.2021.767998

Khalil, M. I., Hossain, M. B., & Schmidhalter, U. (2005). Carbon and nitrogen mineralization in different upland soils of the subtropics treated with organic materials. *Soil Biology and Biochemistry, 37*, 1507–1518. https://doi.org/10.1016/j.soilbio.2005.01.014

Kirihetti, D. (2018). *Direct assessment of the fixation and release of nitrogen by pea, lentil, chickpea and Faba Bean. A master of science thesis*. University of Saskatchewan Saskatoon, Canada.

Kocira, A., Staniak, M., Tomaszewska, M., Kornas, R., Cymerman, J., Panasiewicz, K., & Lipińska, H. (2020). Legume cover crops as one of the elements of strategic weed management and soil quality improvement. A review. *Agriculture, 10*, 394. https://doi.org/10.3390/agriculture10090394

Kohmann, M. M., Sollenberger, L. E., Dubeux, J. C., Jr., Silveira, M. L., & Moreno, L. S. (2019). Legume proportion in grassland litter affects decomposition dynamics and nutrient mineralization. *Agronomy Journal, 111*(3), 1079–1089. https://doi.org/10.2134/agronj2018.09.0603

Kokkini, M., Gazoulis, I., Danaskos, M., Kontogeorgou, V., Kanatas, P., & Travlos, I. (2025). Enhancing ecosystem services in agriculture: The special role of legume intercropping. *Frontiers in Sustainable Food Systems, 9*, 1547879. https://doi.org/10.3389/fsufs.2025.1547879

Lal, R. (2017). Improving soil health and human protein nutrition by pulses-based cropping systems. *Advances in Agronomy, 145*, 167–204. https://doi.org/10.1016/bs.agron.2017.05.003

Lavelle, P., Spain, A. V., Blouin, M., Brown, G. G., Decaëns, T., Grimaldi, M., Jiménez, J. J., McKey, D., Mathieu, J., Velasquez, E., & Zangerlé, A. (2016). Ecosystem engineers in a self-organized soil: A review of concepts and future research questions. *Soil Science, 181*(3–4), 91–109.

Lazali, M., & Drevon, J. J. (2023). Legume ecosystemic services in agro-ecosystems: a review communication in plant science. *Communication in Plant Science, 13*, 13–19. https://doi.org/10.26814/cps2023002

Lehmann, J., & Kleber, M. (2015). The contentious nature of soil organic matter. *Nature, 528*, 60–68.

Li, C., Hoffland, E., Kuyper, T. W., Yu, Y., Zhang, C., Li, H., et al. (2020). Syndromes of production in intercropping impact yield gains. *Nature Plants, 6*, 653–660.

Liang, S. (2013). *Short-term effects of legume winter cover crop management on soil microbial activity and particulate organic matter. (Thesis)* (p. 89). North Carolina State University.

Liu, S., Dong, A., Niu, B., Xu, F., Xu, J., et al. (2025a). Maize/cover crop intercropping mitigates soil erosion and enhances yield of ridge cultivation in Chinese Mollisol region. *Catena, 255*, 109012. https://doi.org/10.1016/j.catena.2025.109012

Liu, H., Gao, X., Li, C., Cai, Y., Song, X., & Zhao, X. (2025b). Intercropping increases plant water availability and water use efficiency: A synthesis. *Agriculture, Ecosystems & Environment, 379*, 109360. https://doi.org/10.1016/j.agee.2024.109360

Liu, Y., Wang, B., Wang, Y., Gao, Q., Yao, F., Wu, H., & Sun, X. (2025c). Intercropping outweighs straw incorporation driving community and functional diversity of microarthropods after 5 years of tillage practices. *Agriculture, Ecosystems & Environment, 377*, 109251. https://doi.org/10.1016/j.agee.2024.109251

Luo, L., Guo, M., Wang, E., Yin, C., Wang, Y., He, H., & Zhao, C. (2022). Effects of mycorrhiza and hyphae on the response of soil microbial community to warming in eastern Tibetan plateau. *Science of the Total Environment, 837*, 155498.

Makoi, J. H., Bambara, S., & Ndakidemi, P. A. (2013). *Rhizobium* inoculation and the supply of molybdenum and lime affect the uptake of macro elements in common bean (*P. vulgaris L.)* plants. *Australian Journal of Crop Science, 7*, 784–793.

Masson-Boivin, C., & Sachs, J. L. (2018). Symbiotic nitrogen fixation by rhizobia—The roots of a success story. *Current Opinion in Plant Biology, 44*, 7–15. https://doi.org/10.1016/j.pbi.2017.12.001

Matse, D. T., Huang, C., Huang, Y., & Yen, M. (2020). Effects of co-inoculation of *Rhizobium* with plant growth promoting rhizobacteria on the nitrogen fixation and nutrient uptake of *Trifolium repens* in low phosphorus soil. *Journal of Plant Nutrition, 43*, 739–752. https://doi.org/10.1080/01904167.2019.1702205

Meng, Y., Li, P., Xiao, L., Hu, B., Zhang, C., Yang, S., Liu, J., & Zhao, B. (2024). Differences in dissolved organic matter and analysis of influencing factors between plantations pure and mixed forest soils in the loess plateau. *Frontiers in Forests and Global Change, 7*, 1344784. https://doi.org/10.3389/ffgc.2024.1344784

Milcu, A., Partsch, S., Langel, R., & Scheu, S. (2008a). The response of decomposers (earthworms and microorganisms) to variations in species and functional diversity of plant communities. *Oikos, 117*(5), 691–700.

Milcu, A., Partsch, S., Scherber, C., Weisser, W. W., & Scheu, S. (2008b). Earthworms and legumes control litter decomposition in a plant diversity gradient. *Ecology, 89*(7), 1872–1882. https://doi.org/10.1890/07-1377.1. PMID: 18705374.

Mmbaga, G. W., Mtei, K. M., & Ndakidemi, P. A. (2014). Extrapolations on the use of rhizobium inoculants supplemented with phosphorus (P) and potassium (K) on growth and nutrition of legumes. *Agricultural Sciences, 5*, 1207–1226. https://doi.org/10.4236/as.2014.512130

Nair, A., & Ngouajio, M. (2012). Soil microbial biomass, functional microbial diversity, and nematode community structure as affected by cover crops and compost in an organic vegetable production system. *Applied Soil Ecology, 58*, 45–55.

Naresh, R. K., Timsina, J., Bhaskar, V., Gupta, R. K., Singh, A. K., Dhaliwa, S. S., et al. (2017). Effects of tillage, residue and nutrient management on soil organic carbon dynamics and its fractions, soil aggregate stability and soil carbon sequestration: A review. *EC Nutrition, 12*(2), 53–80.

Nazir, M. J., Li, G., Nazir, M. M., Zulfiqar, F., Siddique, K. H. M., Iqbal, B., & Du, D. (2024). Harnessing soil carbon sequestration to address climate change challenges in agriculture. *Soil and Tillage Research, 237*, 105959. https://doi.org/10.1016/j.still.2023.105959

Nyawade, S. O., Karanja, N. N., Gachene, C. K. K., Gitari, H. I., Schulte-Geldermann, E., & Parker, M. (2019). Short-term dynamics of soil organic matter fractions and microbial activity in smallholder legume intercropping systems. *Applied Soil Ecology, 142*, 123–135.

Osunde, A., Tsado, P., Bala, A., & Sanginga, N. (2004). Productivity of a maize-promiscuous soybean intercrop as affected by fertilizer in the Southern Guinea savanna zone of Nigeria. *West African Journal of Applied Ecology, 5*, 51–62. https://doi.org/10.4314/wajae.v5i1.45598

Ouda, S., & Zohry, A. (2023). Increasing sustainability by intercropping legume crops with sugar beet under imposed water stress. *Irrigation and Drainage*, 1–26. https://doi.org/10.1002/ird.2884

Ouda, S., & Zohry, A. (2024). Increasing land and water use efficiencies of wheat: Case study of Egypt. In *Integration of legume crops with cereal crops under changing climate: Sustainably increases food production*. Springer Publishing House. ISBN: 978-3-031-68101-1.

Ouda, S., Zohry, A., & Abdel-Wahab, S. (2020). Sustainable intensive cropping to reduce irrigation-induced erosion: II. Changing cropping sequence under sprinkler irrigation practice. *Moroccan Journal of Agricultural Science., 1*(1), 26–33.

Palm, C., Blanco-Canqui, H., DeClerck, F., Gatere, L., & Grace, P. (2014). Conservation agriculture and ecosystem services: An overview. *Agriculture, Ecosystems & Environment, 187*, 87–105. https://doi.org/10.1016/j.agee.2013.10.010

Perera, P. C. D., Gruss, I., Twardowski, J., et al. (2022). The impact of restoration methods for *Solidago*-invaded land on soil invertebrates. *Scientific Reports, 12*, 16634. https://doi.org/10.1038/s41598-022-20812-5

Philippot, L., Chenu, C., Kappler, A., Rillig, M. C., & Fierer, N. (2024). The interplay between microbial communities and soil properties. *Nat. Revista de Microbiologia, 22*, 226–239. https://doi.org/10.1038/s41579-023-00980-5

Power, A. G. (2010). Ecosystem services and agriculture: Tradeoffs and synergies. *Philosophical Transactions of the Royal Society B: Biological Sciences, 365*, 2959–2971.

Rodríguez, B. C., Zuazo, V. H. D., Rodríguez, M. S., García-Tejero, I. F., Ruiz, B. G., et al. (2022). Legumes protect the soil erosion and ecosystem services. In *Advances in legumes for sustainable intensification* (pp. 247–266). Academic Press. https://doi.org/10.1016/B978-0-323-85797-0.00010-0

Rojas-downing, M. M., Nejadhashemi, A. P., Harrigan, T., & Woznicki, S. A. (2017). Climate risk management climate change and livestock: Impacts, adaptation, and mitigation. *Climate Risk Management, xxx*, 1–19. https://doi.org/10.1016/j.crm.2017.02.001

Schmidt, O. (2002). Why do cereal–legume intercrops support large earthworm populations? *Applied Soil Ecology, 19*(2), 181–190.

Shanmugam, S., Hefner, M., Pelck, J. S., Labouriau, R., & Kristensen, H. L. (2022). Complementary resource use in intercropped faba bean and cabbage by increased root growth and nitrogen use in organic production. *Soil Use and Management, 38*, 729–740. https://doi.org/10.1111/sum.12765

Sileshi, G., & Mafongoya, P. L. (2006). Long-term effects of improved legume fallows on soil invertebrate macrofauna and maize yield in eastern Zambia. *Agriculture, Ecosystems & Environment, 115*(1–4), 69–78. https://doi.org/10.1016/j.agee.2005.12.010

Sissoko, A., & Kpomblekou, A. K. (2010). Carbon decomposition in boiler litter-treated soils soil biol. *The Biochemist, 42*, 543–550. https://doi.org/10.1016/j.soilbio.2009.10.007

Smýkal, P., Coyne, C. J., Ambrose, M. J., Maxted, N., Schaefer, H., Blair, M. W., et al. (2014). Legume crops phylogeny and genetic diversity for science and breeding. *Critical Reviews in Plant Sciences, 34*(1–3), 43–104. https://doi.org/10.1080/07352689.2014.897904

Stagnari, F., Maggio, A., Galieni, A., et al. (2017). Multiple benefits of legumes for agriculture sustainability: An overview. *Chemical and Biological Technologies in Agriculture, 4*, 2. https://doi.org/10.1186/s40538-016-0085-1

Taha, A., & Kasem, E. (2022). Water management of maize-cowpea intercropping system under surface irrigation. *Moroccan Journal of Agricultural Sciences, 3*(1), 29–38.

Tamta, A., Kumar, R., Ram, H., Meena, R. K., Kumar, U., Yadav, M. R., et al. (2019). Nutritional portfolio of maize and cowpea fodder under various intercropping ratio and balanced nitrogen fertilization. *The Indian Journal of Animal Sciences, 89*, 276–280. https://doi.org/10.56093/ijans.v89i3.88041

Tang, X., Zhang, C., Yu, Y., Shen, J., van der Werf, W., & Zhang, F. (2021). Intercropping legumes and cereals increases phosphorus use efficiency; a meta-analysis. *Plant and Soil, 460*, 89–104. https://doi.org/10.1007/s11104-020-04768-x

Thakur, M. P., Tilman, D., Purschke, O., Ciobanu, M., Cowles, J., Isbell, F., Wragg, P. D., & Eisenhauer, N. (2015). Plant diversity drives soil microbial biomass carbon in grasslands irrespective of global environmental change factors. *Global Change Biology, 21*(11), 4076–4085.

Tuma, J., Fleiss, S., Eggleton, P., Frouz, J., Klimes, P., Lewis, O. T., Yusah, K. M., & Fayle, T. M. (2019). Logging of rainforest and conversion to oil palm reduces bioturbator diversity but not levels of bioturbation. *Applied Soil Ecology, 144*, 123–133.

Wang, L., Chen, X., Cheng, B., Yang, H., Gao, Y., Li, X., et al. (2025). Stage-specific phosphorus mobilization enhances phosphorus uptake in relay-intercropped soybean. *Agriculture, Ecosystems & Environment, 388*, 109660. https://doi.org/10.1016/j.agee.2025.109660

Wardle, D. A., Bardgett, R. D., Klironomos, J. N., Setälä, H., van der Putten, W. H., & Wall, D. H. (2004). Ecological linkages between aboveground and belowground biota. *Science, 304*(5677), 1629–1633.

Winck, B., Chauvat, M., Coulibaly, S., Santonja, M., de Sa, E. L. S., & Forey, E. (2020). Functional collembolan assemblages induce different plant responses in Lolium perenne. *Plant and Soil, 452*, 347–358.

Wortman, S. E., Francis, C. A., & Lindquist, J. L. (2012). Cover crop mixtures for the western Corn Belt: Opportunities for increased productivity and stability. *Agronomy Journal, 104*, 699–705.

Xiao, X., Cheng, Z., Meng, H., Liu, L., Li, H., & Dong, Y. (2013). Intercropping of green garlic (Allium sativum L.) induces nutrient concentration changes in the soil and plants in continuously cropped cucumber (Cucumis sativus L.) in a plastic tunnel. *PLoS One, 8*(4). https://doi.org/10.1371/journal.pone.0062173

Yuvaraj, M., Pandiyan, M., & Gayathri, B. (2020). Role of legumes in improving soil fertility status. In *Legume crops—Prospects, production and uses*. Intechopen. https://doi.org/10.5772/intechopen.93247

Zhang, K., Liu, Z., McCarl, B. A., & Fei, C. J. (2024). Enhancing agricultural soil carbon sequestration: A review with some research needs. *Climate, 12*, 151. https://doi.org/10.3390/cli12100151

Zohry, A., & Ouda, S. (2025). Legume crops enhance water use efficiency under intercropping system with wheat. *Moroccan Journal of Agricultural Sciences, 6*(3), 145–160.

Zohry, A., Ouda, S., & Abdel-Wahab, T. (2020). Sustainable intensive cropping to reduce irrigation-induced erosion: I. Intercropping systems under surface irrigation practice. *Moroccan Journal of Agricultural Science, 1*(2), 63–71.

Chapter 4
The Agroecosystem Service of Legumes in Climate Regulation

Abstract Legumes offer essential agroecosystem functions by modulating microclimates and improving resilience to weather fluctuations and climate change. This chapter analyzes the impact of legume-based intercropping systems on field microclimatic parameters, such as light intensity, canopy temperature, and relative humidity, and their consequent influence on water use efficiency and water equivalent ratio. Evidence indicates that legume intercropping enhances light distribution via canopy stratification, alleviates heat stress by regulating canopy and soil temperatures, and maintains humidity inside the canopy. These microclimate modifications enhance agricultural yield under current and future climate. Legumes improve soil physical characteristics, organic matter levels, and water infiltration, therefore augmenting soil water retention and resilience to adverse weather conditions. The water equivalent ratio legume-based combinations often attain values exceeding one under both full and deficit irrigation conditions. Case studies from Egypt demonstrate that legume intercropping systems routinely enhance water use efficiency, achieving improvements between 22% and 30% with field crops, and up to 80% when combined with perennial systems like orange orchards. Under water-limited situations, water use efficiency improvements of 13–29% have been shown, highlighting the capacity of legumes to mitigate agricultural production challenges posed by drought stress. Furthermore, a simulation study in Egypt on the effect of future climate scenarios revealed improvement in water productivity of maize and wheat when two legume crops were included in the cropping sequence, compared to the cropping sequence that included maize and wheat only. The findings collectively confirm the strategic importance of legumes in sustainable agriculture by improving microclimate regulation, resource efficiency, and climate resiliency.

Keywords Legume-based intercropping systems · Light intensity · Canopy temperature · Relative humidity · Water use efficiency · Water equivalent ratio · Future climate scenarios

A. E. Zohry, S. A. H. Ouda, *The Agroecosystem Services of Leguminous Crops*,
https://doi.org/10.1007/978-3-032-32461-0_4

Introduction

Agroecosystems are human-managed systems designed to sustain food production, yet they are increasingly challenged by soil degradation, biodiversity loss, and climate variability (Banerjee et al., 2020). Beyond their provisioning role, these systems also deliver important regulatory ecosystem services, among which climate regulation is particularly significant (Akchaya et al., 2025). This function operates through multiple biophysical processes. For instance, as described by Kumar et al. (2017), crop transpiration and evaporation from soil and irrigation water contribute to atmospheric moisture fluxes, which can moderate local and regional climates. These processes help buffer temperature extremes, reduce heat stress, and influence microclimatic conditions within agricultural landscapes. In addition, vegetation cover in agroecosystems enhances energy balance by increasing latent heat flux and reducing sensible heat, thereby mitigating the intensity of heat fluctuations (Lin et al., 2022). Collectively, these mechanisms highlight the critical role of well-managed agroecosystems not only in food production but also in stabilizing climatic conditions and enhancing environmental resilience.

Weather is the primary factor that regulates season duration and, consequently, crop development and final production. The World Meteorological Organization (2012) defines weather as the short-term atmospheric conditions. Global warming and environmental pollution have elevated the average global surface temperature and diminished solar radiation, consequently increasing the incidence of environmental stresses, such as heat and cold stresses, which negatively impact crop growth and yield (Ambavaram et al., 2014). The frequency and scale of extreme weather events are anticipated to intensify due to the expected climate change (Solomon et al., 2007). Seasonal weather fluctuations can lead to numerous issues, including elevated atmospheric carbon dioxide levels, heightened daytime and nighttime temperatures, and irregular rainfall patterns (Meuwissen et al., 2019). Exposure to extreme weather events may exacerbate infections of insects, pests, and plant diseases (John et al., 2024). They may potentially affect soil processes and nutrient dynamics (Miralles et al., 2019) and markedly affect soil moisture by elevating evaporative demand during heatwaves, resulting in soil water depletion (Furtak & Wolińska, 2023).

The intensification of spatio-temporal climate variability under ongoing climate change is projected to profoundly disrupt agroecosystem functioning, thereby diminishing their capacity to sustain the provision of essential goods and ecosystem services and heightening their vulnerability to degradation (Rositano et al., 2018). Climate variability refers to any alteration in climate that may be attributed to natural phenomena or human activities (Hegerl et al., 2000). The phrase "climate variability" commonly refers to fluctuations in climate statistics within a brief timeframe, such as a particular month, season, or year (Sami Ul Din et al., 2019). Weather variability inherently results in extreme weather events, which are more atypical and severe than normal conditions, including heat waves (high-temperature events), cold waves (low-temperature events), downpours (heavy-precipitation events), and droughts (low-precipitation events) (van der Wiel & Bintanja, 2021).

There is a substantial body of evidence that demonstrates a persistent alteration in global climatic conditions (Ali & Mujeeb-Kazi, 2021). Climate change is linked to human activities that alter the composition of the global atmosphere and contribute to observed climate variability across similar time frames (Karmakar et al., 2016). Global climate models forecast an increase in atmospheric carbon dioxide concentration, as well as rising temperatures, altered surface solar radiation, and fluctuating precipitation (Knox et al., 2016). A rise by 1–6 °C in the average ambient temperature is expected by the end of the twenty-first century (De Costa, 2011).

The hydro-physical properties of soil are considerably influenced by variations in climatic parameters caused by climate change, such as seasonal temperature and precipitation intensity, which affect soil water dynamics. Climate change-induced water scarcity resulting from elevated temperatures is expected to negatively impact crop development (Gitari et al., 2020). Projected rises in world average temperature resulting from climate change may adversely affect soil fertility and quality by exacerbating evapotranspiration losses (Mandal & Roy, 2024). An elevation in evapotranspiration rates may augment the salinity of the root zone by enhancing the capillary movement of water and solutes from groundwater (Várallyay, 2010). Cole et al. (2018) demonstrated that elevated temperatures exacerbate soil CO_2 emissions, resulting in increased atmospheric temperatures.

Legumes can mitigate climate variability by improving soil health and bolstering the resilience of agricultural systems to changing environmental conditions (Rahman et al., 2022). Legumes exhibit many unique characteristics, namely morphological, physiological, and biochemical characteristics that provide high resilience and adaptability to changing climatic conditions (Snapp et al., 2019). Intercropping with legumes can enhance climate regulation by modifying the microclimate inside the field, influencing aspects such as light intensity, temperature (Akchaya et al., 2025), humidity, desiccating winds, and other climatic elements (Gebru, 2015). Moreover, the breakdown of legume residues releases nitrogen and other elements that improve soil fertility for subsequent crops, hence alleviating yield losses induced by environmental stresses (Srivastava et al., 2025). Their ability to improve soil health also strengthens resilience by enhancing water retention and alleviating the impacts of extreme weather events (Vadez et al., 2012). Many legume species have drought resistance and can withstand temperature fluctuations, making them beneficial in regions experiencing increased variability in precipitation and temperature (Dutta et al., 2022). Legumes improve soil structure by fostering aggregation and water retention, so diminishing soil erosion and enhancing water infiltration, which bolsters the soil's resilience to drought and excessive rainfall (Rahman et al., 2022).

Crop diversification approaches augment net crop output and promote soil health by promoting interspecific interactions among crop species both aboveground and belowground (Rakshit et al., 2018). Aboveground diversification increases the canopy's capacity to absorb heat and light, whereas belowground diversification boosts the efficiency of water and soil macro- and micro-nutrient use by stimulating soil microbial activity (King & Hofmockel, 2017). Legume-based intercropping systems can mitigate the adverse effects of climate change by altering the microclimate through strategically designed intercropping arrangements (Gitari et al., 2020).

This chapter examines the function of legumes in regulating microclimatic factors within intercropping systems and evaluates their impact on enhancing water use efficiency and sustainable crop yield in both optimal and water-deficient conditions. The contribution of legumes to enhancing climate resilience was also addressed.

Legume-Based Intercropping Systems Modify Microclimate

Legumes Intercropping Systems Alter Light Intensity

The inclusion of legumes as intercrops alters light penetration, allowing shade-tolerant plants to thrive in the shade created by taller species, while taller plants also provide shade for shorter ones (Nyawade et al., 2019a). Intercropping enhances the complementarity of light utilization per unit area, contingent upon climatic alterations facilitated by crop mixes with varying rooting capacities, canopy configurations, heights, and nutrient demands (Nyawade et al., 2019b). Furthermore, intercropping enhances light utilization efficiency per unit area by modifying the microclimate through the integration of crop combinations with varying rooting capabilities, canopy structures, heights, and nutrient demands (Nyawade et al., 2019b).

C3 intercrops experience partial shading within intercropping systems, which reduces light intensity and enhances their efficiency in utilizing available light compared to monocultures (Harris et al., 1987). Moreover, monoculture agriculture of soybean has been associated with oxidative damage to leaves due to heightened light intensity (Shang et al., 2023); hence, intercropping soybean with maize may enhance its output. Maize, characterized by its taller and more upright canopy, can intercept and exploit light more efficiently than soybeans in an intercropping system. The crop's microenvironment differs from the upper to the lower canopy (Mahi & Kingra, 2013). Elevating maize planting density within its intercropping system with soybean diminished light penetration to the lower canopy (Liu et al., 2018), resulting in accelerated withering of the lower leaves (Borrás et al., 2003), and lowered light usage efficiency (Jia et al., 2018). Moreover, the diminished light intensity in soybean and maize intercropping systems has been documented to decrease root exudation and modify root morphology under low light circumstances, likely resulting in a decline in soil nematode populations as light intensity fell (Wu et al., 2024). Intercropping leguminous crops with potatoes enhances light use efficiency relative to the monoculture cultivation of potatoes (Nyawade et al., 2020). The maize-peanut intercropping system improves light use and promotes soil nutrient efficiency compared to monoculture planting (Zhang et al., 2019).

In Egypt, Abdel-Wahab et al. (2024) indicate that intercropping soybeans with maize reduced nematode and azotobacter populations compared to monoculture soybean cultivation, where nematode behavior and reproduction are influenced by light intensity. Abdel-Wahab and Abd El-Rahman (2016) asserted that maize and soybean plants vie for sunlight within an intercropping system at varying maize plant densities, influencing light use efficiency. The light use efficiency in soybean and maize intercropping systems is affected by the disparities in canopy architecture and photosynthetic capability between the two crops. Intercropping maize with soybean at a ratio of 2:2 maize to soybean or 2:4 maize to soybean showed enhanced efficiency in light energy utilization (Abdel-Wahab et al., 2019). Abdel-Wahab and El Manzlawy (2016a) intercropped faba bean (2 rows) with wheat at three planting densities (6, 5, and 4 rows), resulting in increased light intensity in the center of the wheat canopy as the number of wheat rows decreased from six to four, thereby enhancing light use efficiency. Comparable outcomes were observed in soybean intercropped with sunflower, where decreasing sunflower plant density from 100% to 50% led to an enhancement in light intensity within the soybean canopy's interior (Abdel-Wahab & El Manzlawy, 2016b).

Legumes Intercropping Systems Alter Canopy Temperature

Intercropping with legumes can substantially alter and generally reduce canopy temperature by augmenting the leaf area index and offering denser ground cover, which boosts soil shade and improves soil moisture retention (Shumet et al., 2022). Gebru (2015) elucidated that intercropping can mitigate temperature extremes and reduce temperatures within the canopy's microclimate. Moreover, during heat stress, intercropping facilitates the formation of microclimates that lower soil and canopy temperatures (Nyawade et al., 2019c).

Intercropping systems reduced soil temperature, thereby enhancing soil water content and crop yield (Li, 2016). Furthermore, the canopy shadowing of intercropped plants leads to a drop in soil temperature by diminishing radiation and the heating effect on the soil surface, hence decreasing evapotranspiration (Ai et al., 2021). The decrease in air temperature beneath maize canopies was found to promote the growth of groundnut (Awal et al., 2006). Furthermore, according to Lin et al. (2020), a reduction in ambient canopy temperature within the peanut canopy during nighttime and before and after noon, in comparison to monoculture, with a maximum recorded difference of 4.9 °C under its intercropping system with maize. Molla et al. (2023) report a 2–3 °C reduction in canopy temperatures and a 10% increase in yield under heat stress in a cowpea-maize intercropping system. Murphy et al. (2021) similarly discovered that intercropping with pigeon pea produced a 15% higher yield compared to monocultures in elevated temperatures. Intercropping green gram with sorghum reduced soil temperature by 7.2 °C during the growing season (Yumbya et al., 2024). Moreover, intercropping faba bean with wheat and the use of nitrogen fertilizer decreased canopy temperature in comparison to the values observed during monoculture farming (Ma et al., 2023). Likewise, intercropping legumes with potatoes enhanced the leaf area index by 26–57% compared to monoculture cultivation of potato and considerably reduced soil temperatures at a depth of 0–30 cm by up to 7.3 °C, while increasing soil water content by up to 38% (Nyawade et al., 2019c).

The temperature alteration by legumes is significantly evident when faba beans are intercropped with tomatoes in Egypt throughout the winter. Farmers in Egypt commonly utilize faba bean intercropping with tomatoes to potentially safeguard tomatoes from cold and wind damage during winter. This intercropping system results in increased tomato yield, enhanced land use efficiency, and improved farmer revenue. It also extends the tomato growth season by two months (Ibrahim et al., 2010) and improves water efficiency, hence fostering a more productive and sustainable agricultural framework in the region (Abd El-Aal & Zohry, 2004). The enhanced water use efficiency (yield divided by consumed water) of this system is attributable to the application of the same irrigation volumes to the faba bean–tomato intercropping system, similar to the amount applied to monoculture of tomatoes. Moreover, the production of marketable tomatoes was enhanced with intercropping due to the cold protection afforded by faba bean (Abd El-Aal & Zohry, 2004). Elevated faba bean plants create a protective microclimate for the shorter tomato plants, enhancing air and soil temperatures while diminishing wind exposure during frigid winter months, resulting in decreased damage and improved yields (Abd El-Hady et al., 2013). Ibrahim et al. (2010) reported that intercropping faba beans with tomatoes resulted in a 36% increase in tomato yield compared to monoculture agriculture. Furthermore, Elsadany and El-Shamy (2016) asserted that intercropping faba beans with tomatoes diminishes the insect pest burden on tomatoes and lessens reliance on chemical pest control, resulting in healthier crop production.

Legumes Intercropping Systems Alter Canopy Relative Humidity

Legumes modify canopy relative humidity via transpiration, which emits water vapor from the leaves, and by affecting air circulation inside the canopy (Chia & Lim, 2022). Denser legume growth can retain moisture, elevating humidity, while their photosynthetic processes can influence the environment (Farrel & Altieri, 1995). An overlapping, thicker canopy resulting from intercropping offers greater shade than monocultures, which diminishes air circulation and reduces temperatures inside and surrounding the canopy, thereby lowering evaporation rates and enhancing relative humidity (Shumet et al., 2022). The diverse canopy architecture and light interception from intercropping can impact transpiration and the microclimate, resulting in alterations in humidity that may affect crop production and resource utilization (Akchaya et al., 2025). Moreover, alterations in canopy architecture, light absorption, and water utilization patterns linked to intercropping impact the rates of transpiration and evapotranspiration from both plants and soil, thereby directly influencing relative humidity (Adam et al., 2025).

Reduced temperatures result in a diminished saturation deficit of the air, defined as the disparity between the existing moisture content and the maximal moisture capacity at a given temperature (Blöcher et al., 2023). A reduced deficit indicates that the air is nearer to achieving full moisture saturation, hence enhancing relative humidity (Nyawade et al., 2020). Additionally, the legume canopy can diminish soil moisture loss by evaporation, hence enhancing the overall moisture content of the intercropped system (Gebru, 2015).

Furthermore, intercropping resulted in diminished daytime relative humidity and heightened nighttime relative humidity, hence facilitating reduced water loss from plants during the day and the maintenance of moisture levels at night (He et al., 2011). The modification of relative humidity by legume-based intercropping system frequently elevated during daylight hours, so fostering a more stable and resilient microclimate, particularly in arid circumstances (Gebru, 2015). Legume intercropping generates a shadier and more humid microclimate within the canopy (Barry et al., 2019), as evidenced by maize–legume intercropping (Fu et al., 2023). For example, intercropping maize with peanuts elevated the canopy's relative humidity during daylight, with a maximum reported difference of 21% (Lin et al., 2020).

Although research in Egypt does not explicitly indicate that legumes influence canopy relative humidity, it is conceivable that legume cultivation may affect canopy humidity through factors such as enhanced soil moisture from irrigation linked to leguminous crops, the physical presence of the crops, and possible atmospheric water vapor release; however, specific studies are required to validate these effects within the Egyptian context.

The Consequences of Microclimate Modification on Enhancing Water Use Efficiency

The water consumption of crops is affected by meteorological factors, including solar radiation, temperature, and relative humidity. Lobell et al. (2009) assert that the unpredictability linked to weather patterns significantly contributes to the disparity between potential and actual yield, with estimates indicating a 20% gap in some instances (Zhang et al., 2013). Intercropping systems that include legumes can markedly enhance soil physic-hydraulic parameters and refine soil structure (Shen et al., 2023). Legume intercrops may enhance soil moisture levels by providing shade, which reduces soil temperatures (Nyawade et al., 2019c). Intercropping systems including legumes enhance soil water retention by improving soil structure, augmenting organic matter content, thus boosting the soil's capacity for infiltration and water retention (Akchaya et al., 2025).

Legume roots assist in binding soil particles, forming more stable aggregates and disrupting hardpans, so enhancing water infiltration and diminishing runoff

(Bolinder et al., 2020). Their nitrogen and biomass contributions enhance soil fertility and overall soil health, resulting in improved water conservation (Ghanbari et al., 2010). The legume intercropping system results in a dense canopy, which frequently correlates with decreased soil moisture due to heightened transpiration from the canopy (von Arx et al., 2013), thereby diminishing the likelihood of evaporation (Mao et al., 2012). Ghanbari et al. (2010) assert that the enhancement of water acquisition efficiency in intercropping results from augmented root density in the upper soil strata.

Water Utilization Efficiency in Intercropping Systems

Legume intercrops enhance water use efficiency by affecting soil moisture levels, thereby modifying relative humidity (Brooker et al., 2015). Intercropping legumes with cereals has demonstrated a 25% improvement in water use efficiency compared to monocultures by optimizing water absorption and minimizing soil evaporation, especially under drought situations (Fernández-Ortega et al., 2023). Shen et al. (2024) reported a substantial increase in soil water content while intercropping tiger nuts with legumes. Shen et al. (2023) similarly observed an elevation in soil water content when maize was intercropped with soybean, attributed to enhanced nitrogen fixation and the production of acidic metabolites from the intercropped legumes, resulting in a reduction of soil pH (Liu et al., 2023), which are the primary osmotic pressure regulators, enhancing soil water content (Rezig et al., 2013).

The integration of deep-rooted and shallow-rooted species facilitates the extraction of water resources from deeper soil layers that would otherwise be inaccessible (Li et al., 2013). In a maize–pea intercropping system, regional variations in water uptake between monocultures and intercrops demonstrated complementing root structures (Mao et al., 2012). Furthermore, Mupangwa et al. (2021) indicate that in a maize intercropping system with groundnut, the shallow rooting system of groundnut efficiently utilized surface moisture, thereby reducing competition for subsoil water necessary for maize, resulting in a 25% increase in water use efficiency compared to monoculture maize. Dube et al. (2014) suggested that the advantages of cowpea intercropping with maize may be due to enhanced moisture retention in intercrops relative to sole crops. Furthermore, Dai et al. (2019) emphasized the advantages of sorghum–cowpea intercropping, as cowpea roots mostly utilized surface moisture, enabling sorghum to tap into deeper soil water reserves.

Water Equivalent Ratio

The water equivalent ratio (WER) is a statistic employed in intercropping systems to assess water use efficiency and evaluate the effectiveness of water utilization within the intercropping system (Mao et al., 2012). WER indicates the water use efficiency of intercropping systems and links the yield of intercrops with their water consumption (Chimonyo et al., 2016). A WER value over 1.0 for intercropping systems signifies superior water use efficiency of intercropped species compared to monoculture (Feng et al., 2016). WER is defined as the aggregate water required for sole crops to achieve an equivalent crop production per unit area of intercrop (Mao et al., 2012).

$$\begin{aligned} WER &= \left[\left(Y_{int,A} / WU_{int}\right) / \left(Y_{mono,A} / WU_{mono,A}\right)\right] \\ &+ \left[\left(Y_{int,B} / WU_{int}\right) / \left(Y_{mono,B} / WU_{mono,B}\right)\right] \end{aligned} \quad (4.1)$$

where: $Y_{int,A}$, WU_{int}, $Y_{mono,A}$ and $WU_{mono,A}$ are yield of intercropped crop A, water use by the intercropping system, yield of mono crop A, and water use by mono crop A, respectively. $Y_{int,B}$, $Y_{mono,B}$, $WU_{mono,B}$ are yield of intercropped crop B, yield of mono crop B, and water use by mono crop B, respectively.

Under intercropping, WER greater than 1 indicates that the intercropped system is more water-efficient than sole cropping, meaning less water is required to produce the same amount of intercropped yield compared to individual sole crops (Mao et al., 2012). The favorable outcomes are frequently associated with the enhanced soil coverage offered by the legume crop, which diminishes evaporation, and the legumes' capacity to fix nitrogen, potentially reducing the necessity for supplementary fertilizer and augmenting overall system productivity and water use efficiency (Stagnari et al., 2017). Furthermore, the particular crops cultivated in conjunction can markedly affect the value of WER, with certain combinations demonstrating superior water usage efficiency compared to others (Mao et al., 2012).

In legume-based intercropping system, the improved water use efficiency is often attributed to complementary water use between the legume and the other crop, such as the legume's deep root system and the cereal's shallower roots, as well as improved soil moisture conservation through reduced runoff and evaporation (Oubenali et al., 2024). According to Mao et al. (2012), relay intercropping peas with maize attained a value of WER of 1.16. Similarly, in maize intercropping system with peanut, a WER value of 1.14 attained as an average over a three-year experiment (Sun et al., 2025). Whereas, WER value ranged from 1.17 to 1.22, for two intercropping patterns of peanut intercropping system with millet (Feng et al., 2016). Miri et al. (2024) reported a value of WER equal to 1.46 when mung bean was intercropped with maize. Furthermore, Raza et al. (2022) intercropped soybean with maize in three planting densities and concluded that the highest value of WER reaching 1.55 was found when maize was intercropped in medium planting density. In another study by Raza et al. (2021), the value of WER ranged from 1.32 to 1.49, exhibiting that maize–soybean strip-intercropping is a productive strategy to maximize water use efficiency.

Water Equivalent Ratio of Legume Intercropping Systems in Egypt under Optimal Irrigation Conditions

Research on legume-based intercropping systems in Egypt demonstrates favorable water equivalent ratios (WER), frequently surpassing 1.0, suggesting that these systems exhibit greater water use efficiency compared to monocropping. The cowpea and maize intercropping system can yield a WER of 1.26 (Taha & Kasem, 2022), but the sugar beet and faba bean intercropping system can attain a WER of 1.27 (El-Mehy et al., 2020) under optimal irrigation conditions.

Multiple factors influence WER, including crop combination, planting proportions, density, and growing circumstances. For instance, intercropping 33% of peas with 100% of wheat at their recommended planting density and using the necessary irrigation water resulted in a value of WER of 1.26, indicating a 26% enhancement in water use efficiency. Whereas, intercropping 25% faba bean with 100% wheat at their optimum density and using the necessary irrigation water resulted in a value of WER of 1.22, yielding a 22% enhancement in water usage efficiency. The outcomes are contingent upon the morphology of the co-crop within the intercropping system (Zohry & Ouda, 2025).

The percentage increases in water use efficiency derived from the measured water equivalent ratios for selected intercropping systems reported in Egypt are presented in Table 4.1. These results indicate that intercropping systems incorporating legumes may enhance water use efficiency, achieving values ranging from 22% to 30% for intercropping with field crops and up to 80% for interplanting legumes among orange trees.

Water Equivalent Ratio under Water Stress Conditions

Intercropping systems have been shown to mitigate the impact of abiotic stresses on crop yield, including water stress (Asgharipour & Rafiei, 2010; Layek et al., 2018), with water use efficiency expected to be enhanced in drought conditions (Yin et al., 2020). For example, Zohry and Ouda (2025) conducted an intercropping system of faba bean with wheat in Egypt under moderate (85% ETc) and severe (70% ETc)

Table 4.1 Percentage improvement in water use efficiency (WUE) estimated using various published intercropping methods in Egypt, irrigated with the requisite amount of water

Intercropping system	Increase in WUE (%)	References
50% sunflower+100% peanut	30%	Ouda et al. (2018)
25% faba bean+100% sugar beet	30%	El-Mehy et al. (2020)
50% faba bean+100% orange trees	70%	Zohry et al. (2020)
50% lupine+100% orange trees	80%	Zohry et al. (2020)
25% faba bean+100% sugar beet	14%	Abd Allah et al. (2021)
25% faba bean+100% sugar beet	8%	Ouda and Zohry (2023)
25% chickpea+100% sugar beet	13%	Ouda and Zohry (2023)

water stress circumstances. WER values were 1.18 and 1.11 in response to irrigation with 85% and 70% ETc, respectively, in comparison to the values derived from the necessary irrigation application, where the reduction in applied water was 17% and 35%, respectively. These results could be attributed to the fact that legume crops improve soil structure and promote soil aggregation, hence strengthening water retention, increasing water holding capacity, and improving water infiltration, which augments the soil's tolerance to water stress.

The percentage of increase in water usage efficiency values under water stress, derived from the measured WER values for selected intercropping systems reported in Egypt is presented in Table 4.2. The results in Table 4.2 implied that intercropping with legumes mitigates the detrimental effects of moderate water stress and can positively influence water usage efficiency.

Overcoming Climate Change Setbacks Using Legume-Based Cropping System

Agricultural production is influenced by climate change through both direct and indirect pathways. Rising mean temperatures directly accelerate crop growth and development, while shifts in seasonal precipitation patterns, coupled with increased evaporative demand, indirectly intensify drought stress (IPCC, 2012). Increased temperatures have been shown to shorten crop cycles by hastening developmental and phenological stages (Eyshi Rezaei et al., 2018; Martins et al., 2019), ultimately leading to reduced yields. Meanwhile, changes in precipitation regimes affect water availability, which can severely limit productivity or even result in crop failure (Asseng et al., 2013). Overall, climate change poses significant challenges to agricultural systems, particularly in water-limited environments (Paymard et al., 2018).

Climate change is anticipated to significantly impact global wheat and maize output, with prior research indicating a reduction in their yields in the future. Liu and Yang (2010) asserted that a temperature increase of 2–5 °C in China might adversely impact wheat yield more significantly than an elevation in CO_2

Table 4.2 Percentage improvement in water use efficiency (WUE) derived from various reported intercropping systems in Egypt under moderate water stress

Intercropping system	Increase in WUE (%)	References
50% cowpea + 100% maize	13%	Taha and Kasem (2022)
25% faba bean + 100% sugar beet	29%	Noreldin and Abd-Allah (2022)
33% sesame + 100% peanut	27%	El-Mehy et al. (2023)
25% fahl clover* + 100% wheat	26%	Ouda and Zohry (2024a)
25% faba bean + 100% wheat	18%	Zohry and Ouda (2025)
33% peas + 100% wheat	23%	Zohry and Ouda (2025)
25% fahl clover* + 100% wheat	18%	Zohry and Ouda (2025)

*A variety of Egyptian clover

concentration alone. Liu et al. (2016) reported that for each 1 °C increase in temperature, worldwide wheat yields are anticipated to decrease by 4–6%. Furthermore, Asseng et al. (2015) estimated that an increase in temperature due to climate change will reduce global wheat yield by 6% for every additional 1 °C rise.

Likewise, the impact of climate change on global maize output has been thoroughly examined. Msowoya et al. (2016) stated that rain-fed maize production in Malawi may decrease up to 14% by mid-century due to climate change, rising to as much as 33% loss by the end of the century. Zhao et al. (2017) projected a reduction in maize yield by 7% for every degree increase in global mean temperature. Kogo et al. (2019) projected a decline in maize production by as much as 8–38% under RCP4.5 and RCP8.5 scenarios by the end of the twenty-first century. Whereas, Choruma et al. (2019) projected a decrease in maize production by 24% using climate scenario RCP8.5 resulted from MIROC model in 2070–2099 period.

In a simulation study executed by Ouda and Zohry (2024b) in Egypt, the SALTMED crop model and CMIP6 mean projections for two future scenarios, specifically SSP1-2.6 and SSP5-8.5, across two time-intervals, 2060–2079 and 2080–2099, were utilized to assess the contribution of legumes in bolstering the resilience of maize and wheat to climate change within three cropping sequences. The three examined cropping sequences were maize followed by wheat (MW), maize followed by fahl clover and then wheat (MFW), and cowpea intercropped with maize, followed by fahl clover and then wheat (CMFW). Fahl clover (a variety of Egyptian clover), an early-winter crop, is sown in September and harvested in early November. Their findings indicated that, owing to increased temperatures across all examined future climatic scenarios, the reduction in maize and wheat yields in MW was the most significant. The yield decline of maize and wheat was less pronounced in MFW, with the most minimal decrease observed in CMFW, where the adverse impact of elevated temperatures was mitigated by incorporating two legume crops into the crop sequence: cowpea and fahl clover. Cowpea intercropped with maize generated early ground cover, hence diminishing soil evaporation, whereas maize plants contributed to lowering canopy temperature and fahl clover provided ground cover and enhanced soil organic matter prior to wheat production in mid-November. Additionally, the inclusion of fahl clover in MFW crop sequence, and the inclusion of cowpea and fahl clover in CMFW crop sequence enhanced the water productivity (yield per unit of applied water) of wheat and maize under the studied scenarios of future climate, compared to its value obtained under MW crop sequence. Consequently, CMFW demonstrated greater resilience to anticipated elevated temperatures in the future.

Legume-Based Cropping Systems Can Mitigate Water Stress Induced by Climate Change

Climate change-induced water deficit adversely affects crop development, although these impacts can be mitigated through well-designed legume intercropping systems (Gitari et al., 2020).

The anticipated climate change will elevate air temperatures, resulting in increased soil evaporation losses and less moisture availability for cultivated crops (Mandal & Roy, 2024). In this context, studies by Blum (2011), Yang et al. (2011), Rahman et al. (2017), Chen et al. (2018), and others have shown that intercropping systems have many advantages. This encompasses minimizing runoff, enhancing soil moisture utilization, and augmenting productivity per unit of irrigation water applied, ultimately resulting in increased water usage efficiency in the context of expected climate change. According to Makumba et al. (2009), intercropping can modify specific plant characteristics related to water acquisition, including root distribution, as well as the depth of water sourcing (Sekiya & Yano, 2004). Specific investigations indicated that a legume-based intercropping system with maize cultivated under water stress yielded results comparable to or exceeding those of sole maize (Steward et al., 2019).

Ouda et al. (2020) assert that when the provided irrigation water is equivalent to that used for sole maize cultivation, intercropping cowpeas with maize yields a compensatory soil water effect. Maize plants access deeper soil layers for water, whereas cowpea plants utilize surface layers. Post-harvest of cowpea, intercropped maize persisted in utilizing the soil moisture typically allocated to cowpea. Consequently, the production of cowpeas matched that of its monoculture, whereas the yield of maize exceeded it. Intercropping systems enhance water usage efficiency by diminishing soil evaporation and augmenting crop transpiration and soil water content via early foliage cover and increased leaf area (Maitra et al., 2023; Rezig et al., 2013). This results in a more resilient vegetation cover, hence diminishing soil evaporation. Research by Chimonyo et al. (2016) indicates that intercropping cowpeas and sorghums enhances water use efficiency by 55% relative to monoculture. Research by Choudhary and Kumar (2016) reported that intercropping maize and legumes enhances water use efficiency by 15–55% relative to monoculture planting.

Generally, these evidences clearly demonstrates that legume-based intercropping systems offer an effective strategy for mitigating water stress under climate change conditions.

Conclusion

Legume-based intercropping systems offer a comprehensive method for managing agroecosystem microclimates and enhancing resource use efficiency. Legumes boost yield stability, mitigate abiotic stress, and improve soil health via altering light interception, canopy temperature, and relative humidity. Their contributions to water conservation and efficiency, measured by enhanced WUE and WER values, are especially important in arid and semi-arid locations like Egypt, where water shortage is a considerable challenge. Furthermore, in times of water stress, legumes exhibit beneficial impacts on WUE, underscoring its adaptive superiority in climate-resilient agricultural systems. In the context of future climate change, the incorporation of legumes into cropping systems can significantly enhance the resilience of cultivated crops.

These findings underscore the significance of integrating legumes into varied cropping systems as a sustainable approach to enhance resource utilization, alleviate the effects of climate unpredictability, and bolster food security.

References

Abd Allah, A., Mohamed, T. S., Mohamed, M. S., & Noreldin, T. (2021). Alleviation of water deficiency effect by application of potassium silicate to faba bean intercropped with sugar beet in sandy soil. *Zagazig Journal of Agricultural Research, 48*(1), 1–18.

Abd El-Aal, A. I. N., & Zohry, A. A. (2004). Land and water exploitation for maximizing productivity of tomato intercropping with faba bean crop under stress ecological conditions at Toshky. *Egypt, Journal of Agricultural Research, 82*(1), 37–51.

Abd El-Hady, M. A., Marey, R. A., & Abou-Keriasha, M. A. (2013). Protection of tomato plants against cold and frost conditions by intercropping under Sohag governorate. *Journal of Plant Production, Mansoura University, 4*(7), 1077–1096.

Abdel-Wahab, T. I., & Abd El-Rahman, R. A. (2016). Response of some soybean cultivars to low light intensity under different intercropping patterns with maize. *International Journal of Applied Agricultural Sciences., 2*(2), 21–31. https://doi.org/10.11648/j.ijaas.20160202.11

Abdel-Wahab, S. I., & El Manzlawy, A. M. (2016a). Yield and quality of intercropped sunflower with soybean under different sunflower plant spacings and slow—Release nitrogen fertilizer rates in sandy soil. *International Journal of Applied Agricultural Sciences, 2*(3), 32–43.

Abdel-Wahab, T. I., & El Manzlawy, A. M. (2016b). Yield and quality of intercropped wheat with faba bean under different wheat plant densities and slow—Release nitrogen fertilizer rates in sandy soil. *American Journal of Experimental Agriculture, 11*(6), 1–22.

Abdel-Wahab, T. I., Abdel-Wahab, S. I., & Abdel-Wahab, E. I. (2019). Benefits of intercropping legumes with cereals. *International Journal of Conference Proceedings, 1*(2), ICP.000510.2019.

Abdel-Wahab, T. I., Abdel-Wahab, S. I., & Abdel-Wahab, E. I. (2024). Biological engineering and its relationship to nematode resistance (Chapter 15, pp: 383–408). In K. K. Chaudhary, M. K. Meghvansi, & S. Siddiqui (Eds.), *Sustainable management of nematodes in agriculture, Vol. 2: Role of microbes—assisted strategies, sustainability in plant and crop protection*. Springer Nature.

Adam, A. M., Giller, K. E., Rusinamhodzi, L., Rasche, F., Koomson, E., et al. (2025). Enhancing the resilience of intercropping systems to changing moisture conditions in Africa through the integration of grain legumes: A meta-analysis. *Field Crops Research, 321*, 109663. https://doi.org/10.1016/j.fcr.2024.109663

Ai, P., Ma, Y., & Hai, Y. (2021). Influence of jujube/cotton intercropping on soil temperature and crop evapotranspiration in an arid area. *Agricultural Water Management, 256*(June), 107118. https://doi.org/10.1016/j.agwat.2021.107118

Akchaya, K., Parasuraman, P., Pandian, K., Vijayakumar, S., Thirukumaran, K., Mustaffa, M. R. A. F., Rajpoot, S. K., & Choudhary, A. K. (2025). Boosting resource use efficiency, soil fertility, food security, ecosystem services, and climate resilience with legume intercropping: A review. *Frontiers in Sustainable Food Systems, 9*, 1527256. https://doi.org/10.3389/fsufs.2025.1527256

Ali, N., & Mujeeb-Kazi, A. (2021). Food production: Global challenges to mitigate climate change. In *Physiological, molecular, and genetic perspectives of wheat improvement* (pp. 1–13). Springer. https://doi.org/10.1007/978-3-030-59577-7_1

Ambavaram, M. M. R., Basu, S., Krishnan, A., Ramegowda, V., Batlang, U., Rahman, L., Baisakh, N., & Pereira, A. (2014). Coordinated regulation of photosynthesis in rice increases yield and tolerance to environmental stress. *Nature, 36*, 93–93.

Asgharipour, M., & Rafiei, M. (2010). Intercropping of Isabgol (*Plantago ovata* L.) and Lentil as influenced by drought stress. *American-Eurasian Journal of Sustainable Agriculture, 4*(3), 341–348.

Asseng, S., Ewert, F., Rosenzweig, C., Jones, J. W., & Hatfield, J. L. (2013). Uncertainty in simulating wheat yields under climate change. *Nature Climate Change, 3*, 827–832.

Asseng, S., Ewert, F., Martre, P., Rotter, R. P., Lobell, D., Cammarano, D., Kimball, B., Ottman, M., Wall, G., & White, J. W. (2015). Rising temperatures reduce global wheat production. *Nature Climate Change, 5*, 143–147. https://doi.org/10.1038/nclimate2470

Awal, M. A., Koshi, H., & Ikeda, T. (2006). Radiation interception and use by maize/peanut intercrop canopy. *Agricultural and Forest Meteorology, 139*(1–2), 74–83. https://doi.org/10.1016/j.agrformet.2006.06.001

Banerjee, A., Jhariya, M. K., Yadav, D. K., & Raj, A. (2020). *Environmental and sustainable development through forestry and other resources*. Apple Academic Press, CRC Press- a Tayler.

Barry, K. E., Mommer, L., van Ruijven, J., Wirth, C., Wright, A. J., et al. (2019). The future of complementarity: Disentangling causes from consequences. *Trends in Ecology & Evolution, 34*(2), 167–180. https://doi.org/10.1016/j.tree.2018.10.013

Blöcher, J. R., Diamantopoulos, E., Durner, W., & Iden, S. C. (2023). Validating coupled flow theory for bare-soil evaporation under different boundary conditions. *Vadose Zone Journal*. https://doi.org/10.1002/vzj2.20277

Blum, A. (2011). *Plant breeding for water-limited environment*. Springer. ISBN 978-1-4419-7490-7.

Bolinder, M. A., Crotty, F., Elsen, A., et al. (2020). The effect of crop residues, cover crops, manures and nitrogen fertilization on soil organic carbon changes in agroecosystems: A synthesis of reviews. *Mitigation and Adaptation Strategies for Global Change, 25*, 929–952. https://doi.org/10.1007/s11027-020-09916-3

Borrás, L., Maddonni, G. A., & Otegui, M. E. (2003). Leaf senescence in maize hybrids: Plant population: Row spacing and kernel set effects. *Field Crops Research, 82*(1), 13–26.

Brooker, R. W., Bennett, A. E., & Cong, W.-F. (2015). Improving intercropping: A synthesis of research in agronomy, plant physiology and ecology. *New Phytologist*. https://doi.org/10.1111/nph.13132

Chen, G., Kong, X., & al Gan, Y. e. (2018). Enhancing the systems productivity and water use efficiency through coordinated soil water sharing and compensation in strip-intercropping. *Scientific Reports, 8*, 10494. https://doi.org/10.1038/s41598-018-28612-6

Chia, S. Y., & Lim, M. W. (2022). A critical review on the influence of humidity for plant growth forecasting. In *33rd symposium of Malaysian chemical engineers (SOMChE 2022) IOP Conf. Series: Materials science and engineering 1257 012001*. IOP Publishing. https://doi.org/10.1088/1757-899X/1257/1/012001

Chimonyo, V. G. P., Modi, A. T., & Mabhaudhi, T. (2016). Assessment of sorghum-cowpea intercrop system under water-limited conditions using a decision support tool. *Water SA, 42*(2), 316–327. https://doi.org/10.4314/wsa.v42i2.15

Choruma, D. J., Balkovic, J., & Odume, O. N. (2019). Calibration and validation of the EPIC model for maize production in the Eastern Cape, South Africa. *Agronomy, 9*(9), 494.

Choudhary, V. K., & Kumar, P. S. (2016). Productivity, water use and energy profitability of staggered maize–legume inter-cropping in the Eastern Himalayan Region of India. *Proceedings of the National Academy of Sciences, India Section B: Biological Sciences, 86*(3), 547–557.

Cole, M. B., Augustin, M. A., Robertson, M. J., & Manners, J. M. (2018). The science of food security. *npj Science of Food, 2*, 14. https://doi.org/10.1038/s41538-018-0021-9

Dai, J., Qiu, W., Wang, N., Wang, T., Nakanishi, H., & Zuo, Y. (2019). From Leguminosae/Gramineae intercropping systems to see benefits of intercropping on iron nutrition. *Frontiers in Plant Science, 10*, 605. https://doi.org/10.3389/fpls.2019.00605

De Costa, W. A. J. M. (2011). A review of the possible impacts of climate change on forests in the humid tropics. *Journal of the National Science Foundation of Sri Lanka, 39*, 281–302. https://doi.org/10.4038/jnsfsr.v39i4.3879

Dube, E. D. N., Madanzi, T., Kapenzi, A., & Masvaya, E. (2014). Root length density in maize/cowpea intercropping under a basin tillage system in a semi-arid area of Zimbabwe. *American Journal of Plant Sciences, 5*, 1499–1507. https://doi.org/10.4236/ajps.2014.511165

Dutta, A., Trivedi, A., Nath, C. P., Gupta, D. S., & Hazra, K. K. (2022). A comprehensive review on grain legumes as climate-smart crops: Challenges and prospects. *Environmental Challenges, 7*, 100479. https://doi.org/10.1016/j.envc.2022.100479

El-Mehy, A., Abd Allah, A. M., Mohamed, T. S., & Kasem, E. E. (2020). Intercropping of some faba bean cultivars with sugar beet using different irrigation intervals under sprinkler system in sandy soils of plant production. *Journal of Plant Production., 11*(12), 1215–1225.

El-Mehy, A. A., Abd-Allah, A. M., Kasem, E. E., & Mohamed, M. S. (2023). Mitigation of the impact of water deficiency on intercropped sesame and peanut systems through foliar application of potassium-silicate and triacontanol. *Egyptian Journal of Agricultural Research, 101*(2), 292–303. https://doi.org/10.21608/ejar.2023.191179.1327

Elsadany, M. F. I., & El-Shamy, M. A. (2016). Influence of faba bean-tomato intercropping on mites, insects infestation and the relationship between the pests and predators, leaf phenols content, and yield components of both crops in different intercropping systems. *Mansoura Journal of Biology*, 55–68.

Eyshi Rezaei, E., Siebert, S., Hüging, H., & Ewert, F. (2018). Climate change effect on wheat phenology depends on cultivar change. *Scientific Reports, 8*, 4891. https://doi.org/10.1038/s41598-018-23101-2

Farrel, J. G., & Altieri, M. A. (1995). *The science of sustainable agriculture*. Intermediate Technology Publications. Agroforestry systems in agro-ecology; pp. 219–263.

Feng, L., Sun, Z., Zheng, M., Muchoki, M., Zheng, J., Yang, N., Bai, W., Feng, C. H., Zhang, Z., Cai, Q., & Zhang, D. (2016). Productivity enhancement and water use efficiency of peanut-millet intercropping. *Pakistan Journal of Botany, 48*(4), 1459–1466.

Fernández-Ortega, J., Álvaro-Fuentes, J., Talukder, R., Lampurlanés, J., & Cantero-Martínez, C. (2023). The use of double-cropping in combination with no-tillage and optimized nitrogen fertilization improve crop yield and water use efficiency under irrigated conditions. *Field Crops Research, 301*, 109017. https://doi.org/10.1016/j.fcr.2023.109017

Fu, Z., Chen, P., Zhang, X., Du, Q., Zheng, B., Yang, H., Luo, K., Lin, P., Li, Y., Pu, T., Wu, Y., Wang, X., Yang, F., Liu, W., Song, C., Yang, W., & Yong, T. (2023). Maize-legume intercropping achieves yield advantages by improving leaf functions and dry matter partition. *BMC Plant Biology, 23*(1), 438. https://doi.org/10.1186/s12870-023-04408-3

Furtak, K., & Wolińska, A. (2023). The impact of extreme weather events as a consequence of climate change on the soil moisture and on the quality of the soil environment and agriculture—A review. *Catena, 231*, 107378. https://doi.org/10.1016/j.catena.2023.107378

Gebru, H. A. (2015). Review on the comparative advantage of intercropping systems. *Journal of Biology, Agriculture and Healthcare, 6*, 2.

Ghanbari, A., Dahmardeh, M., Siahsar, B. A., & Ramroudi, M. (2010). Effect of maize (Zea mays L.) - cowpea (Vigna unguiculata L.) intercropping on light distribution, soil temperature and soil moisture in arid environment. *Journal of Food, Agriculture & Environment, 8*(1), 102–108.

Gitari, H. I., Nyawade, S., Kamau, S., Raza, M. A., Karanja, N. N., & Gachene, C. K. K. (2020). Revisiting intercropping indices with respect to potato-legume intercropping systems. *Field Crops Research, 258*, Article 107957.

Harris, D., Natarajan, M., & Willey, R. W. (1987). Physiological basis for yield advantage in a sorghum/groundnut intercrop exposed to drought. 1. Dry-matter production, yield, and light interception. *Field Crops Research, 17*(3–4), 259–272.

He, H., Yang, L., Fan, L., Zhao, L., Wu, H., et al. (2011). The effect of intercropping of maize and soybean on microclimate. *5th computer and computing Technologies in Agriculture (CCTA), Beijing, China*. pp. 257–263, ff10.1007/978-3-642-27278-3_27ff.Ffhal-01360987f.

Hegerl, G. C., Zwiers, F. W., Braconnot, P., Gillett, N. P., Luo, Y., Orsini, J. A. M., Nicholls, N., & Hoogenboom, G. (2000). Contribution of agrometeorology to the simulation of crop production and its applications. *Agricultural and Forest Meteorology, 103*, 137–157.

Ibrahim, S., Shaaban, M., & Gendy, E. K. (2010). Intercropping faba bean with tomato. *Egyptian Journal of Applied Sciences, 25*(6A), 167–181.

IPCC. (2012). In C. B. Field, V. Barros, T. F. stocker, D. Qin, D. J. Dokken, K. L. Ebi, M. D. Mastrandrea, K. J. Mach, G.-K. Plattner, S. K. Allen, M. Tignor, & P. M. Midgley (Eds.), *Managing the risks of extreme events and disasters to advance climate change adaptation. A special report of working groups I and II of the intergovernmental panel on climate change* (p. 582). Cambridge University Press.

Jia, Q., Sun, L., Mou, H., Ali, S., Liu, D., Zhang, Y., Zhang, P., Ren, X., & Jia, Z. (2018). Effects of planting patterns and sowing densities on grain filling, radiation use efficiency and yield of maize (*Zea mays* L.) in semi-arid regions. *Agricultural Water Management, 201*, 287–298.

John, A., Riat, A. K., Bhat, K. A., Ganie, S. A., et al. (2024). Adapting to climate extremes: Implications for insect populations and sustainable solutions. *Journal for Nature Conservation, 79*, 126602. https://doi.org/10.1016/j.jnc.2024.126602

Karmakar, R., Das, I., Dutta, D., & Rakshit, A. (2016). Potential effects of climate change on soil properties: A review. *Science International, 4*, 51–73.

King, A. E., & Hofmockel, K. S. (2017). Diversified cropping systems support greater microbial cycling and retention of carbon and nitrogen. *Agriculture, Ecosystems and Environment, 240*, 66–76.

Knox, J., Daccache, A., Hess, T., & Haro, D. (2016). Meta-analysis of climate impacts and uncertainty on crop yields in Europe. *Environmental Research Letters, 11*, 113004.

Kogo, B. K., Kumar, L., Koech, R., & Langat, P. (2019). Modelling impacts of climate change on maize (Zea mays L.) growth and productivity: A review of models, outputs and limitations. *Journal of Geoscience and Environment Protection, 7*, 76–95.

Kumar, R., Mishra, V., Buzan, J., Kumar, R., Shindell, D., & Huber, M. (2017). Dominant control of agriculture and irrigation on urban heat island in India. *Scientific Reports, 7*, 14054.

Layek J., A. Das, T. Mitran, C. Nath, R.S. Meena, G.S. Yadav, et al., 2018. Cereal+legume intercropping: An option for improving productivity and sustaining soil health R.S. Meena, A. Das, G.S. Yadav, R. Lal (Eds.), Legumes for soil health and sustainable management, Springer, , Singapore, pp. 347–386, https://doi.org/10.1007/978-981-13-0253-4_11.

Li, L. (2016). Intercropping enhances agroecosystem services and functioning, current knowledge and perspectives. *Chinese Journal of Eco-Agriculture, 24*, 403–415. (In Chinese).

Li, L., Zhang, L., & Zhang, F. (2013). Crop mixtures and the mechanisms of overyielding. In S. A. Levin (Ed.), *Encyclopedia of biodiversity (second edition)* (pp. 382–395). Academic Press. https://doi.org/10.1016/B978-0-12-384719-5.00363-4

Lin, S., Meng, W., Nan, Z., Xu, J., Li, L., Zhang, Z., & Li, X. (2020). Canopy microenvironment change of peanut intercropped with maize and its correlation with pod yield. *Chinese Journal of Eco-Agriculture, 28*(1), 31–41. https://doi.org/10.13930/j.cnki.cjea.190525

Lin, H., Li, Y., & Zhao, L. (2022). Partitioning of sensible and latent heat fluxes in different vegetation types and their spatiotemporal variations based on 203 FLUXNET sites. *Journal of Geophysical Research: Atmospheres, 127*, e2022JD037142. https://doi.org/10.1029/2022JD037142

Liu, J., & Yang, H. (2010). Spatially explicit assessment of global consumptive water uses in cropland: Green and blue water. *Journal of Hydrology, 384*, 187–197.

Liu, B., Asseng, S., Muller, C., Ewert, F., & Elliott, J. (2016). Similar estimates of temperature impacts on global wheat yield by three independent methods. *Nature Climate Change*, 1130–1136.

Liu, X., Rahman, T., Song, C., Yang, F., Su, B., Cui, L., Bu, W., & Yang, W. (2018). Relationships among light distribution, radiation use efficiency and land equivalent ratio in maize-soybean strip intercropping. *Field Crops Research, 224*, 91–101.

Liu, H. Y., Huang, N., Zhao, C. M., & Li, J. H. (2023). Responses of carbon cycling and soil organic carbon content to nitrogen addition in grasslands globally. *Soil Biology and Biochemistry, 186*, 109164. https://doi.org/10.1016/j.soilbio.2023.109164

Lobell, D. B., Cassman, K. G., & Field, C. B. (2009). Crop yield gaps: Their importance, magnitudes, and causes. *Annual Review of Environment and Resources, 34*, 179–204.

Ma, G., Zheng, Y., Zhang, J., Guo, Z., & Dong, Y. (2023). Changes in canopy microclimate of faba bean under intercropping at controlled nitrogen levels and their correlation with crop yield. *Journal of the Science of Food and Agriculture, 103*(9), 4489–4502. https://doi.org/10.1002/jsfa.12533

Mahi, G. S., & Kingra, P. K. (2013). *Comprehensive agrometeorology* (p. 355). Kalyani Publishers.

Maitra, S., Sahoo, U., Sairam, M., Gitari, H. I., Rezaei-Chiyaneh, E., Battaglia, M. L., & Hossain, A. (2023). Cultivating sustainability: A comprehensive review on intercropping in a changing climate. *Research on Crops, 24*(4), 702–715. https://doi.org/10.31830/2348-7542.2023.ROC1020

Makumba, W., Akinnifesi, F. K., & Janssen, B. H. (2009). Spatial rooting patterns of gliricidia, pigeon pea and maize intercrops and effect on profile soil N and P distribution in southern Malawi. *African Journal of Agricultural Research, 4*, 278–288. https://doi.org/10.5897/AJAR.9000652

Mandal, D., & Roy, T. (2024). Climate change impact on soil erosion and land degradation. In H. Pathak, D. Chatterjee, S. Saha, & B. Das (Eds.), *Climate change impacts on soil-plant-atmosphere continuum. Advances in global change research* (Vol. 78). Springer. https://doi.org/10.1007/978-981-99-7935-6

Mao, L. L., Zhang, L. Z., Li, W. W., Werf, W. V. D., Sun, J. H., Spiertz, H., & Li, L. (2012). Yield advantage and water saving in maize/pea intercrop. *Field Crops Research, 138*, 11–20.

Martins, M. A., Tomasella, J., & Dias, C. G. (2019). Maize yield under a changing climate in the Brazilian northeast: Impacts and adaptation. *Agricultural Water Management., 216*, 339–350.

Meuwissen, M. P. M., Feindt, P. H., Spiegel, A., Termeer, C. J., Mathijs, E., de Mey, Y., Finger, R., Balmann, A., Wauters, E., Urquhart, J., et al. (2019). A framework to assess the resilience of farming systems. *Agricultural Systems, 176*.

Miralles, D. G., Gentine, P., Seneviratne, S. I., & Teuling, A. J. (2019). Land-atmospheric feedbacks during droughts and heatwaves: State of the science and current challenges. *Annals of the New York Academy of Sciences, 1436*(1), 19–35. https://doi.org/10.1111/nyas.13912

Miri, S., Alizadeh, H. A., Alizadeh, Y., & Amini, E. (2024). Water use efficiency of maize-mung bean in intercropping systems under different water stress conditions. *Water and Soil Management and Modeling, 4*(1), 233–247. https://doi.org/10.22098/mmws.2023.12317.1224

Molla, M. S. H., Kumdee, O., Worathongchai, N., Khongchiu, P., Ali, M. A., Anwar, M. M., et al. (2023). Efforts to stimulate Morpho-physio-biochemical traits of maize for efficient production under drought stress in tropics field. *Agronomy, 13*, 2673. https://doi.org/10.3390/agronomy13112673

Msowoya, K., Madani, K., Davtalab, R., Mirchi, A., & Lund, J. R. (2016). Climate change impacts on maize production in the warm heart of Africa. *Water Resources Management, 30*, 5299–5312. https://doi.org/10.1007/s11269-016-1487-3

Mupangwa, W., Nyagumbo, I., Liben, F., Chipindu, L., Craufurd, P., & Mkuhlani, S. (2021). Maize yields from rotation and intercropping systems with different legumes under conservation agriculture in contrasting agro-ecologies. *Agriculture, Ecosystems and Environment, 306*, 107170. https://doi.org/10.1016/j.agee.2020.107170

Murphy, K. M., Shukla, R., & Kumar, S. (2021). The role of pigeon pea in intercropping systems under high-temperature conditions. *Journal of Crop Science and Biotechnology, 24*, 265–275.

Noreldin, T., & Abd-Allah, A. M. M. (2022). Potassium silicate for mitigation of irrigation water deficiency for faba bean intercropped with sugar beet in a sandy soil. *Moroccan Journal of Agricultural Sciences, 3*(3), 139–147.

Nyawade, S. O., Gachene, C. K. K., Karanja, N. N., Gitari, H. I., Schulte-Geldermann, E., & Parker, M. L. (2019a). Controlling soil erosion in smallholder potato farming systems using legume intercrops. *Geoderma Regional, 15*, e00225.

Nyawade, S. O., Karanja, N. N., Gachene, C. K. K., Gitari, H. I., Schulte-Geldermann, E., & Parker, M. (2019b). Short-term dynamics of soil organic matter fractions and microbial activity in smallholder legume intercropping systems. *Applied Soil Ecology, 142*, 123–135.

Nyawade, S. O., Karanja, N. N., Gachene, C. K. K., et al. (2019c). Intercropping optimizes soil temperature and increases crop water productivity and radiation use efficiency of Rainfed potato. *American Journal of Potato Research, 96*, 457–471. https://doi.org/10.1007/s12230-019-09737-4

Nyawade, S., Gitari, H. I., Karanja, N. N., Gachene, C. K. K., Schulte-Geldermann, E., Sharma, K., & Parker, M. L. (2020). Enhancing climate resilience of rain-fed potato through legume intercropping and silicon application. *Frontiers in Sustainable Food Systems, 4*, 566345. https://doi.org/10.3389/fsufs.2020.566345

Oubenali, A., El Moukhtari, A., Lamsaadi, N., et al. (2024). The potential role of durum wheat–chickpea intercropping in promoting plant growth and yield in low soil phosphorus availability under field conditions. *Euro-Mediterranean Journal for Environmental Integration, 9*, 1325–1345. https://doi.org/10.1007/s41207-024-00515-1

Ouda, S., & Zohry, A. (2023). Increasing sustainability by intercropping legume crops with sugar beet under imposed water stress. *Irrigation and Drainage*, 1–26. Available from:. https://doi.org/10.1002/ird.2884

Ouda, S., & Zohry, A. (2024a). Increasing land and water use efficiencies of wheat: Case study of Egypt. In *Integration of legume crops with cereal crops under changing climate: Sustainably increases food production*. Springer Publishing House. ISBN: 978-3-031-68101-1.

Ouda, S., & Zohry, A. (2024b). Legumes improve wheat and maize productivity grown in different cropping systems under changing climate. In *Integration of legume crops with cereal crops under changing climate: Sustainably increases food production*. Springer Publishing House. ISBN: 978-3-031-68101-1.

Ouda, S., Hefny, Y. A. A., Abdel-Wahab, T. I., & Abdel-Wahab, S. I. (2018). Intercropping systems of sunflower and peanut under different irrigation regimes and potassium fertilizer levels. *Egypt. J. Agron. Proced. 15th Int. Conf. Crop Sci.*, pp. 85–104.

Ouda, S., Zohry, A., & Abdel-Wahab, S. (2020). Sustainable intensive cropping to reduce irrigation-induced erosion: II. Changing cropping sequence under sprinkler irrigation practice. *Moroccan Journal of Agricultural Science., 1*(1), 26–33.

Paymard, P., Bannayan, M., & Haghighi, R. S. (2018). Analysis of the climate change effect on wheat production systems and investigate the potential of management strategies. *Natural Hazards, 91*, 1237. https://doi.org/10.1007/s11069-018-3180-8

Rahman, T., Liu, X., Hussain, S., Ahmed, S., Chen, G., & Yang, F. (2017). Water use efficiency and evapotranspiration in maize-soybean relay strip intercrop systems as affected by planting geometries. *PLoS One, 12*(6), e0178332. https://doi.org/10.1371/journal.pone.0178332

Rahman, M., Alam, M. S., Islam, M., Kamal, M. Z. U., Rahman, G. K. M. M., et al. (2022). Potential of legume-based cropping systems for climate change adaptation and mitigation. In *Advances in legumes for sustainable intensification* (pp. 381–402). Academic Press. https://doi.org/10.1016/B978-0-323-85797-0.00030-6

Rakshit, A., Sarkar, B., & Abhilash, P. C. (2018). *Soil amendments for sustainability: Challenges and perspectives*. CRC Press.

Raza, M. A., Gul, H., Wang, J., Yasin, H. S., Qin, R., Bin Khalid, M. H., Naeem, M., et al. (2021). Land productivity and water use efficiency of maize-soybean strip intercropping systems in semi-arid areas: A case study in Punjab Province, Pakistan. *Journal of Cleaner Production, 308*, 127282. https://doi.org/10.1016/j.jclepro.2021.127282

Raza, M. A., Yasin, H. S., Gul, H., Qin, R., Mohi Ud Din, A., Khalid, M. H. B., Hussain, S., Gitari, H., Saeed, A., Wang, J., Rezaei-Chiyaneh, E., Sabagh, A. E., Manzoor, A., Fatima, A., Ahmad, S., Yang, F., Skalicky, M., & Yang, W. (2022). Maize/soybean strip intercropping produces higher crop yields and saves water under semi-arid conditions. *Frontiers in Plant Science, 13*, 1006720. https://doi.org/10.3389/fpls.2022.1006720

Rezig, M., Sahli, A., Ben Jeddi, F., & Harbaoui, Y. (2013). Potato (Solanum tuberosum L.) and sulla (Hedysarum coronarium L.) intercropping in Tunisia: Effects in water consumption and water use efficiency. *Journal of Arid Sciences, 5*. https://doi.org/10.5539/jas.v5n10p123

Rositano, F., Bert, F. E., Piñeiro, G., & Ferraro, D. O. (2018). Identifying the factors that determine ecosystem services provision in Pampean agroecosystems (Argentina) using a data-mining approach. *Environmental Development, 25*, 3–11. https://doi.org/10.1016/j.envdev.2017.11.003

Sami Ul Din, M., Ahmad, I., Hussain, N., Ahmad, A., Wajid, A., Khaliq, T., Mubeen, M., Imran, M., Ali, A., Akram, R., Amanet, K., Saleem, M., & Nasim, W. (2019). Agronomic cropping systems in relation to climatic variability. In M. Hasanuzzaman (Ed.), *Agronomic crops*. Springer Nature Singapore Pte Ltd. https://doi.org/10.1007/978-981-32-9151-5_5

Sekiya, N., & Yano, K. (2004). Do pigeon pea and sesbania supply groundwater to intercropped maize through hydraulic lift?—Hydrogen stable isotope investigation of xylem waters. *Field Crops Research, 86*, 167–173. https://doi.org/10.1016/j.fcr.2003.08.007

Shang, J., Zhao, L., Yang, X., Qi, X., Yu, J., Du, J., Li, K., He, C., Wang, W., & Yang, W. (2023). Soybean balanced the growth and defense in response to SMV infection under different light intensities. *Frontiers in Plant Science, 14*, 1150870.

Shen, L., Wang, X., Liu, T., Wei, W., Zhang, S., Keyhani, A., et al. (2023). Border row effects on the distribution of root and soil resources in maize–soybean strip intercropping systems. *Soil and Tillage Research, 233*, 105812. https://doi.org/10.1016/j.still.2023.105812

Shen, X., Liu, Y., Li, X., & Li, L. (2024). Effect of intercropping with legumes at different rates on the yield and soil physicochemical properties of Cyperus esculentus L. in arid land. *Frontiers in Plant Science, 15*, 1351843. https://doi.org/10.3389/fpls.2024.1351843

Shumet, S. T., Ayalew, T., Roro, A. G., & Beshir, H. M. (2022). Intercropping and *Rhizobium* inoculation affected microclimate and performance of common bean (*Phaseolus vulgaris L.*) varieties. *Scientifica*, 3471912. https://doi.org/10.1155/2022/3471912

Snapp, S. S., Cox, C. M., & Peter, B. G. (2019). Multipurpose legumes for smallholders in sub-Saharan Africa: Identification of promising 'scale out' options. *Global Food Security, 23*, 22–32. https://doi.org/10.1016/j.gfs.2019.03.002

Solomon, S., Qin, D., Manning, M., Marquis, M., Averyt, K., Tignor, M. M. B., Miller, L. H., & Chen, Z. (Eds.). (2007). *Climate change (2007): The physical science basis. Contribution of working group I to the fourth assessment report of the intergovermental panel on climate change*. Cambridge University Press.

Srivastava, R. K., Yetgin, A., & Srivastava, S. (2025). The role of legume roots in carbon sequestration, soil health enhancement, and salinity mitigation under climate change: A comprehensive review. *Soil and Tillage Research, 253*, 106656. https://doi.org/10.1016/j.still.2025.106656

Stagnari, F., Maggio, A., Galieni, A., et al. (2017). Multiple benefits of legumes for agriculture sustainability: An overview. *Chemical and Biological Technologies in Agriculture, 4*(2). https://doi.org/10.1186/s40538-016-0085-1

Steward, P. R., Thierfelder, C., Dougill, A. J., & Ligowe, I. (2019). Conservation agriculture enhances resistance of maize to climate stress in a Malawian medium-term trial. *Agriculture, Ecosystems and Environment, 277*, 95–104. https://doi.org/10.1016/j.agee.2018.07.009

Sun, X., Zhang, Y., Zhang, S., Yang, N., Xia, G., & Feng, L. (2025). Effect of strip width in maize/peanut intercropping on water use efficiency. *Frontiers in Sustainable Food Systems, 9*, 1502362. https://doi.org/10.3389/fsufs.2025.1502362

Taha, A., & Kasem, E. (2022). Water management of maize-cowpea intercropping system under surface irrigation. *Moroccan Journal of Agricultural Sciences, 3*(1), 29–38.

Vadez, V., Berger, J. D., Warkentin, T., Asseng, S., Ratnakumar, P., Rao, K. P. C., & Sharma, H. C. (2012). Adaptation of grain legumes to climate change: A review. *Agronomy for Sustainable Development, 132*, 31–44. https://doi.org/10.1007/s13593-011-0020-6

van der Wiel, K., & Bintanja, R. (2021). Contribution of climatic changes in mean and variability to monthly temperature and precipitation extremes. *Communications Earth and Environment*. https://doi.org/10.1038/s43247-020-00077-4

Várallyay, G. (2010). The impact of climate change on soils and on their water management. *Agronomy Research, 8*(Special issue II), 385–396.

von Arx, G., Graf Pannatier, E., Thimonier, A., & Rebetez, M. (2013). Microclimate in forests with varying leaf area index and soil moisture: Potential implications for seedling establishment in a changing climate. *Journal of Ecology, 101*, 1201–1213. https://doi.org/10.1111/1365-2745.12121

World Meteorological Organization. (2012). *Standardized precipitation index user guide*. World Meteorological Organization.

Wu, B., Jia, X., Zhu, W., Gao, Y., Tan, K., Duan, Y., Chen, L., Fan, H., Wang, Y., Liu, X., Xuan, Y., & Zhu, X. (2024). Light signaling regulates root-knot nematode infection and development via HY5-SWEET signaling. *BMC Plant Biology, 24*(1), 664.

Yang, C., Huang, G., Chai, Q., & Luo, Z. (2011). Water use and yield of wheat/maize intercropping under alternate irrigation in the oasis field of Northwest China. *Field Crops Research, 124*, 426–432.

Yin, W., Chai, Q., Zhao, C., Yu, A., Fan, Z., Hu, F., Fan, H., Guo, Y., & Coulter, J. A. (2020). Water utilization in intercropping: A review. *Agricultural Water Management, 241*, 106335. https://doi.org/10.1016/j.agwat.2020.106335

Yumbya, B. M., Kitonyo, O. M., & Kinama, J. M. (2024). Green gram-sorghum intercropping: Effects on crop microclimate, competition, resource use efficiency and yield in southeastern Kenya. *International Journal of Environment and Climate Change, 14*(12), 549–565.

Zhang, X. Y., Wang, S. F., Sun, H. Y., Chen, S. Y., Shao, L. W., & Liu, X. W. (2013). Contribution of cultivar, fertilizer and weather to yield variation of winter wheat over three decades: A case study in the North China plain. *European Journal of Agronomy, 50*, 52–59.

Zhang, X. N., Chen, P., Du, Q., Zhou, Y., Ren, J. R., Jin, F., Yang, W. Y., & Yong, T. W. (2019). Effects of maize/soybean and maize/peanut intercropping systems on crops nitrogen uptake and nodulation nitrogen fixation. *Chinese Journal of Eco-Agriculture, 27*, 1183–1194. (In Chinese).

Zhao, C. B., Liu, S. P., Wang, X., Lobell, D. B., Huang, Y., Huang, M., Yao, Y., Bassu, S., Ciais, P., Durand, J.-L., Elliott, J., Ewert, F., Janssens, I. A., Li, T., Lin, E., Liu, Q., Martre, P., Müller, C., Peng, S., … Asseng, S.. (2017). Temperature increase reduces global yields of major crops in four independent estimates. *Proceedings of the National Academy of Sciences of the United States of America 114* (35):9326–9331. https://doi.org/10.1073/pnas.1701762114.

Zohry, A., & Ouda, S. (2025). Legume crops enhance water use efficiency under intercropping system with wheat. *Moroccan Journal of Agricultural Sciences, 6*(3), 145–160.

Zohry, A., Hefny, Y., & Ouda, S. (2020). Interplanting four legume crops under orange trees using different irrigation water and fertilizer sources in sandy soil. *16th International Conference on Crop Science. 10th of October. Agron. Dept., Fac. Agric, Cairo, Al-Azhar Univ., Oct. 13, Cairo, Egypt*. pp. 341–346.

Chapter 5
Modification of Tomato Microclimate by Intercropping with Faba Bean: A Field Study

Abstract Tomato production in Egypt is increasingly constrained by climate variability, particularly low winter temperatures and water scarcity, which adversely affect plant growth, fruit set, and yield. Intercropping systems have been proposed as an effective strategy to modify crop microclimate, improve resource-use efficiency, and enhance productivity under such stress conditions. This chapter investigates the role of faba bean–tomato intercropping in modifying the tomato microclimate and mitigating cold and water stress through a two-year field experiment conducted in the sandy soils of Middle Egypt. Three irrigation regimes (100%, 85%, and 70% of crop evapotranspiration (ETc)), and three faba bean planting densities (25%, 50%, and 75% of the recommended rate) intercropped with tomato were evaluated alongside sole cropping systems of both crops. Results demonstrated that increasing faba bean density enhanced tomato productivity, with the highest yields obtained under 75% faba bean density combined with full irrigation application. Intercropping extended the tomato growing season, moderated temperature fluctuations, and reduced cold stress, resulting in higher fruit yield and improved microclimatic conditions. Land equivalent ratio (LER) and water equivalent ratio (WER) values consistently exceeded unity across treatments, indicating superior land and water productivity of the intercropping system relative to sole cropping. Although water stress reduced yields of both crops, intercropping mitigated these effects, maintaining higher productivity and resource-use efficiency than monoculture. The findings confirm that faba bean–tomato intercropping is an effective climate-resilient strategy for improving tomato production and it offers a practical pathway for sustainable intensification in water-limited and climate-vulnerable agroecosystems.

Keywords Faba bean planting densities · Water stress · Land equivalent ratio · Water equivalent ratio · Change in water use

A. E. Zohry, S. A. H. Ouda, *The Agroecosystem Services of Leguminous Crops*,
https://doi.org/10.1007/978-3-032-32461-0_5

Introduction

Tomato (*Solanum lycopersicum* L.) is one of the most economically important vegetable crops in Egypt and a cornerstone of both domestic consumption and agricultural exports. Egypt ranks among the leading global producers of tomato, occupying the fifth position worldwide with an annual production of approximately 7–8 million tons. This production supports large local markets and export demands, particularly for processed tomatoes to European markets and fresh tomatoes to Middle Eastern countries. Despite its economic importance, tomato production in Egypt faces several constraints, including rising input costs, water scarcity, and market instability. Tomatoes are cultivated throughout the year under Egypt's favorable climatic conditions, with major production areas concentrated in North Egypt, particularly in Nubaria city, Sharkia, and Alexandria governorates. However, fluctuations in cultivated area and total production often result in supply instability, while increasing costs of fertilizers and fuel for irrigation reduce profitability (Ali, 2026).

The microclimate surrounding a crop refers to the localized atmospheric conditions—temperature, humidity, wind, and radiation—within and immediately above the plant canopy and soil surface. This microenvironment differs from the general climate due to the influence of vegetation structure, soil properties, and crop management practices, and it plays a critical role in regulating plant growth, water use, disease development, and final yield (Tanny, 2013). Understanding and managing crop microclimate is therefore essential for maximizing productivity. Although adequate solar radiation is necessary for photosynthesis, excessive radiation can induce physiological stress. For instance, tomato fruits exposed to intense direct radiation in the summer may develop sunburn, which significantly reduces marketable yield (El-Mehy & Mohamed, 2018). Many physiological processes in plants are also temperature-dependent, with defined optimal temperature ranges for growth and productivity (Meena et al., 2018). For tomato, the optimal temperature range lies between 12 and 32 °C (Liu et al., 2012). Deviations beyond this range can adversely affect plant performance; both low and high temperatures reduce fruit quality and may increase plant mortality (Sheha et al., 2022). Exposure to temperatures exceeding 32 °C alters physiological processes and induces moderate heat stress, leading to reduced fruit yield (Mesa et al., 2022).

Integrating legumes into tomato intercropping systems is a recognized agricultural strategy for temperature moderation, improving the microclimate around the tomato plants, and boosting overall productivity in Egypt. Intercropping faba bean with winter tomato has been shown to extend the tomato growing season and improve land-use efficiency (Abd El-Aal & Zohry, 2004), thereby supporting more productive and sustainable cropping systems (Ibrahim et al., 2010). Improved water-use efficiency in such systems is largely attributed to the application of similar irrigation volumes to both the intercropping system and to the monocrop, resulting in higher yield per unit of water consumed (Zohry & Ouda, 2025). Intercropping also enhances the production of marketable tomato fruits due to the protective effect of faba bean against cold stress (Abd El-Aal & Zohry, 2004). Taller faba bean plants create a favorable microclimate for the shorter tomato plants by increasing air and soil temperatures and reducing wind exposure during cold winter months, thereby minimizing damage and improving yield (Abd El-Hady et al., 2013).

Planting density of faba bean (*Vicia faba* L.) is a critical factor in optimizing intercropping performance, as it influences the balance between beneficial interactions and interspecific competition. Elsadany and El-Shamy (2016) reported that intercropping tomato with 25% faba bean increased tomato yield by 26% compared with sole cropping, with an LER value of 1.75. Their findings also indicated reduced insect pest pressure and lower dependence on chemical pesticides, contributing to safer and more sustainable crop production. A planting density equivalent to 50% of the recommended faba bean rate has been suggested as suitable for intercropping with tomato to minimize competition while maintaining high productivity (Ibrahim et al., 2010). Under such conditions, faba bean–tomato intercropping increased tomato yield by 36% compared with monoculture, achieving a land equivalent ratio (LER) of 2.21 (Ibrahim et al., 2010). Similarly, increasing faba bean density to 75% of the recommended rate further enhanced tomato yield and provided stronger protection against low temperatures (Abd El-Hady et al., 2013). However, the interactive effects of faba bean planting density and water stress on the productivity and resilience of this intercropping system remain largely unexplored.

Globally, research on faba bean–tomato intercropping remains scarce, with most studies conducted in Egypt and only limited investigations elsewhere. For example, Warren Raffa et al. (2022) evaluated the integration of faba bean into organic tomato systems in Italy and reported improvements in soil nutrient availability, nitrogen balance, and tomato productivity, highlighting the potential of this intercrop under Mediterranean conditions. However, the role of faba bean intercropping in enhancing tomato resilience to climate stress, particularly under water deficit conditions, has not been adequately explored. Therefore, this chapter investigates the modification of the tomato microclimate through intercropping with faba bean under field conditions in the sandy soils of Middle Egypt. It also examines the effects of different faba bean planting densities and imposed water stress treatments on tomato productivity and system performance.

Effect of Cold Stress on Tomato

Tomato is inherently sensitive to low temperatures due to its tropical origin, and exposure to chilling stress can adversely affect nearly all stages of plant development, from seed germination and vegetative growth to reproductive performance and final yield (Yousef & Ali, 2019). When temperatures fall below approximately 12 °C, chilling injury may occur, manifested by chlorosis, membrane damage, and reduced enzymatic activity, which collectively suppress plant growth and physiological efficiency (Wang et al., 2024). Furthermore, cold stress disrupts metabolic homeostasis, impairs cell division and expansion, and interferes with essential physiological processes such as photosynthesis, nutrient uptake, and sap flow, ultimately leading to reduced biomass accumulation and plant vigor (Yousef & Ali, 2019).

During reproductive development, cold stress exerts particularly detrimental effects by disrupting anther development, ovule formation, fertilization, and fruit set. These disturbances often result in malformed flowers and substantial declines in fruit yield (Mesa et al., 2022). Alterations in floral meristem regulation under low-temperature conditions can further impair cell differentiation, contributing to misshapen fruits and reduced fruit quality (Mesa et al., 2022). In addition, chilling stress promotes the accumulation of reactive oxygen species (ROS), which damage cellular membranes and intensify stress symptoms unless neutralized by antioxidant defense mechanisms (Huang et al., 2022).

Effect of Water Stress on Tomato

Under unfavorable growth circumstances, the tomato's development and productivity are significantly restricted due to its high susceptibility to both biotic and abiotic stressors (Raja et al., 2020). One of the most important abiotic factors impacting tomato performance is water deficit, which is frequently brought on by drought or low soil moisture and drastically lowers plant vigor, output, and fruit quality (Putti et al., 2023). Stomatal closure, an early and essential adaptation mechanism to limit water loss via transpiration, is the first of a complex suite of physiological responses triggered by water scarcity (Hatamleh et al., 2022). The plant hormone abscisic acid (ABA), which is produced in response to drought, is primarily responsible for stomatal closure. It instructs guard cells to lower stomatal conductance when water availability is low (Hajihashemi, 2019).

Although stomatal closure aids in water conservation, it also restricts the diffusion of CO_2 into the leaf, which inhibits net photosynthesis and causes the photosynthetic electron transport chain to be overly reduced. If antioxidant mechanisms are unable to efficiently scavenge reactive ROS, this imbalance will lead to oxidative stress (Moreira de Carvalho et al., 2020). Increased activity of antioxidant enzymes like catalase, peroxidases, and superoxide dismutase has been associated with drought stress in tomato. This suggests an activated defense response to prevent ROS accumulation and cellular damage (Petrović et al., 2021).

Carbon assimilation and photosynthetic pigments are also adversely affected by water constraint. Due in part to oxidative degradation and lipid peroxidation triggered by ROS, which lowers light absorption and photosynthetic ability, chlorophyll concentration tends to decrease during drought (Putti et al., 2023). Thus, when compared to well-watered controls, drought-stressed tomato plants exhibit considerably lower levels of important photosynthetic metrics such as net photosynthetic rate, stomatal conductance, and chlorophyll fluorescence efficiency (Liang et al., 2020). Reduced biomass and fruit yield are the results of these constraints, which are exacerbated by declines in the development of leaf area, dry matter accumulation, and carbohydrate production (Putti et al., 2023).

Water stress can also affect fruit quality characteristics like lycopene and total carotenoid content, which are crucial for nutritional value, as well as membrane stability, osmotic adjustment substances (like proline), and secondary metabolites (Wadood et al., 2024). When combined, these reactions highlight how a lack of water causes a series of alterations in tomato physiology, ranging from altered photosynthesis and growth to oxidative stress responses and water regulation, which ultimately results in decreased productivity during extended or severe drought conditions (Putti et al., 2023).

Production of Faba Beans in Egypt

The designated area for faba bean cultivation has markedly diminished, declining from 178,531 hectares in 1991 to 32,532 hectares by 2017, attributed to competition with or preference for other winter crops like wheat (Ministry of Agriculture and Land Reclamation, 2018). Egypt's limited faba bean production frequently leads to substantial import dependence, compounded by considerable losses due to pests, illnesses, and detrimental low-price policies. The cultivation of faba bean in newly reclaimed soils has been established as a policy, alongside its intercropping with other crops to enhance productivity.

Impact of Water Stress on Faba Bean

The faba bean is regarded as one of the most drought-sensitive grain legumes, with its performance under water shortage potentially diminished more severely than that of other legume crops, owing to constraints in water relations and photosynthetic function (Muktadir et al., 2020; Daryanto et al., 2015). Water stress adversely impacts faba bean growth and productivity across the entire crop cycle, with frequent declines in germination, plant height, leaf expansion, chlorophyll content, and photosynthetic efficiency under conditions of inadequate irrigation (Abid et al., 2017). Yield losses are typically most significant when drought transpires during reproductive development, especially during flowering and early pod formation, as water deficiency diminishes pod set, hastens senescence, and ultimately reduces seed yield and total carbohydrate accumulation (Abdelhaleim et al., 2022). In certain instances, seed protein concentration may increase due to concentration effects resulting from diminished biomass production (Muktadir et al., 2020; Alshameri et al., 2021). Consistent with these findings, Ramadan et al. (2025) indicated that utilizing merely 50% of the required irrigation water markedly diminished faba bean growth characteristics, photosynthetic pigments, and yield in comparison to full irrigation (100% of the required amount), underscoring the substantial reliance of faba bean productivity on sufficient water availability in arid and semi-arid environments.

Faba Bean–Tomato Intercropping to Alleviate Cold Stress: A Field Study

In regions such as Middle Egypt, night temperatures frequently fall below the optimal range for tomato cultivation during the winter and early spring seasons, leading to reduced fruit set, yield, and marketable quality (Abd El-Aal & Zohry, 2004). Exposure to temperatures below 10–12 °C during these periods can inhibit vegetative growth, delay flowering, and reduce reproductive success, thereby constraining overall productivity (Yousef & Ali, 2019). To mitigate the adverse impacts of chilling stress, growers can adopt cultural practices that buffer plants against temperature extremes and stabilize the crop microclimate.

Intercropping tomato with suitable companion crops has emerged as an effective strategy for moderating microclimatic conditions and enhancing crop performance under low-temperature stress. For example, intercropping tomato with wheat has been shown to improve fruit set and yield by providing protection against frost and chilling conditions. Field studies demonstrated that wheat–tomato intercropping, when implemented with appropriate sowing dates and seeding rates, increased land equivalent ratios (LER) to as high as 1.80 and enhanced tomato yield compared with sole cropping systems (Sheha et al., 2022). Such intercropping practices not only improve productivity under cold-prone conditions but also enhance resource-use efficiency and farm profitability, making them a promising strategy for climate-resilient tomato production systems.

Methodology

This research comprises a biennial field experiment conducted throughout the successive seasons of 2021/22 and 2022/23 in El-Minia Governorate, Middle Egypt (28.2847°N, 30.5279°E), concentrating on tomato intercropping systems with three planting densities of faba bean under three irrigation regimes. The experimental area has an arid climate marked by frigid winters and sweltering, arid summers (Masoued, 2017).

Data on average daily air temperature, relative humidity, and wind speed for the research seasons were obtained from the weather station located 320 m from the experimental site. The soil at the experimental site is categorized as sandy soil. The soil's physical and chemical parameters were assessed following the methodologies of Klute (1986) and Tan (1996) (Tables 5.1 and 5.2). The irrigation water was sourced from groundwater, with an electrical conductivity of 0.98 dS/m and a pH of 7.72.

The experimental design included a split-plot with three replications. The treatments involved the interaction of three irrigation levels with three intercropping systems, wherein faba bean was intercropped with tomato at three planting densities (25%, 50%, and 75% of the recommended rate), in addition to sole faba bean and sole tomato.

The irrigation treatments comprised the required irrigation amount (RI), which accounted for 100% of crop evapotranspiration (ETc) and functioned as the control treatment, in addition to two imposed water stress treatments: WS1, denoting 85% ETc, and WS2, denoting 70% ETc. The irrigation treatments were assigned to the main plots, while the four cropping systems (three intercropping systems and sole crops) were designated to the sub-plots (Fig. 5.1). The plot area spanned 14.4 m^2 (4.8 × 3.0 m) and comprised four cropping systems. The prior crop of tomato was maize in both growing seasons, sown in May and harvested in early September.

Table 5.1 Main physical properties of soil at the experimental site before cultivation

Soil depth (cm)	Particle size distribution			Texture class	Bulk density (g/cm^3)	Field capacity (%)	Permanent wilting point (%)	Available water (%)
	Sand (%)	Silt (%)	Clay (%)					
0–20	93.40	4.70	2.11	Sandy	1.64	12.76	3.61	9.15
20–40	95.50	3.10	1.50		1.75	11.22	2.89	8.33
40–60	96.10	2.85	0.95		1.69	7.43	2.09	5.34

Table 5.2 Chemical properties and available macronutrients of the soil at the experimental site before cultivation

Soil depth	pH	ECe	Soluble cations (meq/L)				Soluble anions (meq/L)			
(cm)	(1:2.5)	(dS m^{-1})	Ca^{+2}	Mg^{+2}	Na^+	K^+	CO_3^{-2}	HCO_3^-	Cl^-	SO_4^{-2}
0–20	7.68	0.72	2.42	1.08	3.12	0.32	–	2.22	3.44	1.28
20–40	7.62	0.70	2.38	1.02	3.12	0.30	–	2.08	3.48	1.26
40–60	7.45	0.69	2.32	1.01	3.10	0.30	–	2.14	3.46	1.13
Available nutrients (mg/kg)										
N				P_20_5				K_20		
13.2				4.85				60.02		

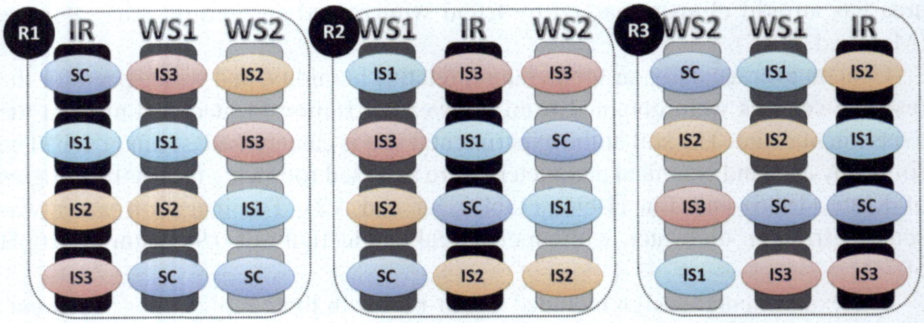

Fig. 5.1 Schematic diagram of the split-plot design used in the experiment. RI, WS_1, and WS_2 = irrigation with 100, 85, and 70% ETc, respectively; IS1, IS2 and IS3 = 25%, 50% and 75% faba bean intercropped with 100% tomato, respectively, SC = sole crop (faba bean or tomato)

Monoculture and Intercropped Tomato

Tomato (cv. Orit improved) transplanting from the nursery occurred on October 1, for both intercropped and sole cultivation. Tomato plants were arranged in the field with a spacing of 40 cm between transplants (one transplant per hill) on the southern side of a raised bed of 1.20 m in width (21,000 plants per hectare), in either a sole or intercropping system. The tomato harvest occurred in late December and January for sole cultivation, whereas it extended until mid-March under intercropping systems.

Calcium superphosphate (15.5% P_2O_5) was administered at a rate of 950 kg/ha during soil preparation in both winter seasons. Nitrogen fertilizer in the form of ammonium sulfate (20.6% N) was applied at a rate of 740 kg/ha in eight installments, commencing 20 days post-transplantation of the tomato, with subsequent applications occurring weekly. Potassium sulfate (48% K_2O) was applied at a rate of 480 kg/ha in four doses. The initial dose was administered prior to transplantation, the second and third doses were given at one-month intervals, and the final dose was administered 2 weeks subsequent to the third dose.

Monocultures and Intercropped Faba Bean

The faba bean was inoculated with *Rhizobium leguminosarum* before sowing, with Arabic gum acting as a binding agent. The cultivar was Giza843, utilized in either monoculture or intercropping regimes. The faba bean seeds were sown on November 5 and 7 during the first and second growing seasons, respectively, and harvested on April 20 and 22 in the first and second growing seasons, respectively (either as a sole crop or intercropped). The planting density of faba bean was 25%, 50%, and 75% of its recommended rate (33,600, 67,200, and 100,800 plants/ha, respectively) in intercropping systems with tomato, and 100% of its recommended rate (134,400 plants/ha) for sole faba bean planting.

At a 25% faba bean density, the seeds were intercropped on the northern sides of the raised beds in a single row, with one plant per hill spaced 25 cm apart. At a 50% faba bean density, the seeds were intercropped on the northern sides of the raised beds in two rows, 20 cm apart, with one plant per hill spaced 25 cm apart. In the 75% faba bean rate, the seeds were intercropped in three rows: two rows were positioned 20 cm apart on the northern sides of the raised beds, while the third row was situated between two raised beds, with one plant per hill spaced 25 cm apart. No fertilizer or irrigation was administered to the intercropped faba bean plants, as they utilize the fertilizer and the water allocated to the tomato plants.

Sole faba bean seeds were sown in four rows atop raised beds, maintaining a spacing of 20 cm between rows and 25 cm between hills, with one plant cultivated per hill. The faba bean was subjected to phosphorus fertilization using single super phosphate (15.5% P_2O_5) at a dose of 600 kg P_2O_5/ha, incorporated into the soil during land preparation. Additionally, during land preparation, potassium was administered in the form of potassium sulfate (48.8% K_2O) at a rate of 120 kg K_2O per hectare. Sole faba bean received 50.0 kg N/ha as ammonium nitrate (33.5% N) in a single application 10 days after seeding.

The computed irrigation water quantities for crop evapotranspiration or irrigation under optimal conditions utilizing a drip system were ascertained using the following equation (Allen et al., 1998):

$$\text{ETc} = \text{ETo x kc} \tag{5.1}$$

where ETc represents crop evapotranspiration (mm/day), ETo denotes reference evapotranspiration (mm/day), and kc signifies the tomato crop coefficient.

ETo values were computed utilizing the Penman-Monteith equation (Allen et al., 1998). The kc values for tomato are 0.94, 1.00, 0.85, and 0.475 for the start, developmental, middle, and end seasons, respectively, as reported by Allen et al. (1998), and were utilized to determine the necessary irrigation water depth.

The requisite irrigation water volume was determined using the subsequent equation (Vermeiren & Jopling, 1984):

$$\text{AIW} = \frac{\text{ETo} \times I}{\text{Ea}\left(1 - \text{LR}\right)} \tag{5.2}$$

AIW represents the applied irrigation water depth (mm/day), whereas Ea denotes the application efficiency, which is 85% for the drip irrigation method. I is the time interval in days and LR is the leaching requirements, which is not considered due to low soil salinity level. We assumed that the irrigation water provided to the intercropped tomato is adequate to fulfill the water needs of its intercropping systems with faba bean. This assumption is supported by the research of Taha and Kasem (2022) for cowpea–maize intercropping system.

Irrigation was performed at scheduled intervals of every 3 days via the drip lateral lines. These lines are linked to the sub-main line. Each lateral line measures 20 m in length and is positioned 0.70 m apart on the sub-main, with integrated emitters with a discharge rate of 2 L/h, spaced 0.30 m along the lateral lines. A differential pressure tank was linked to the drip irrigation system to provide fertilizer through the irrigation water.

The seed yield of faba bean and the fruit output of tomato for all assessed cropping systems under the studied irrigation treatments were quantified by harvesting and weighing all yield of the plants within the experimental plot area, thereafter aggregating the weights of all plots. The biomass of all analyzed crops was harvested from the field after collection. The experiment was performed in the same location used in the first growing season during the second growing season.

Evaluation of the Studied Intercropping Systems

Land Equivalent Ratio (LER)

LER is an evaluation of the land utilization efficiency of the intercropping system (Rao & Willey, 1980) and it is calculated as follows:

$$\mathrm{LER} = \mathrm{LER}_{\mathrm{A}} + \mathrm{LER}_{\mathrm{B}} = \frac{\mathrm{YintA}}{\mathrm{YmonoA}} + \frac{\mathrm{yintB}}{\mathrm{YmonoB}} \tag{5.3}$$

where $\mathrm{LER_A}$ and $\mathrm{LER_B}$ are the partial land equivalent ratios of crop A and crop B, respectively. $Y_{\mathrm{int,A}}$ and $Y_{\mathrm{int,B}}$ are the intercropped yields of crop A and crop B, respectively. $Y_{\mathrm{mono,A}}$ and $Y_{\mathrm{mono,B}}$ are the monoculture yields of crop A and crop B, respectively. If LER exceeds 1, it signifies that the intercropping system demonstrates superior land utilization efficiency compared to monoculture.

Water Equivalent Ratio (WER)

WER measures the volume of water required for individual crops to attain equivalent yields as those produced with one unit of water in intercrops, determined using the formula established by Mao et al. (2012) as follows:

$$\text{WER} = \left[\left(\text{Yint,A / WUint} \right) / \left(\text{Ymono,A / WUmono,A} \right) \right] + \left[\left(\text{Yint,B / WUint} \right) / \left(\text{Ymono,B / WUmono,B} \right) \right] \quad (5.4)$$

where $Y_{int,A}$, $Y_{mono,A}$, $Y_{int,B}$, and $Y_{mono,B}$ are the yields of intercropped crop A, mono crop A, intercropped crop B (legume crop) and mono crop B, respectively. WU_{int}, $WU_{mono,A}$, and $WU_{mono,B}$ are water use by the intercropping system, water use of mono crop A, and water use of intercropped crop B, respectively. If the WER exceeds 1, it indicates that the water output per unit is greater than that of the monoculture.

Change in Water Use (ΔWU)

The ΔWU index (Morris & Garrity, 1993) was utilized in conjunction with WER to assess the advantages of water-use efficiency in intercrops compared to monoculture. ΔWU quantifies the relative discrepancy between the actual water intake in the intercrop (WUint,obs) and the expected water use based on the water consumption of the two crop species, modified by weights that represent their percentage in the intercrop (WUint, exp). The expected water consumption in intercropping is related to the water usage of each species in monoculture and the relative yield attained in intercropping versus monoculture (partial LER). Thus, this formula uses the partial land equivalent ratio of crops A and B as coefficients for ΔWU (Mao et al., 2012) as outlined below:

$$\Delta WU = \left[\text{WUint,obs / WUint,exp} \right] - 1 \quad (5.5)$$

$$\Delta WU = \left[\begin{array}{l} \text{WUint / } \left(\text{LERint,A x WUmono,A} \right) \\ + \left(\text{LERint,B x WUmono,B} \right) \end{array} \right] - 1 \quad (5.6)$$

Both WER and ΔWU assess whether a specific intercrop yield (comprising two species) would be attained with increased water (WER > 1; ΔWU < 0) or decreased water (WER < 1; ΔWU > 0) compared to that utilized in sole crops.

Statistical Analysis

Data from each growing season were assessed separately using a split-plot design with three replications. Statistical analysis was conducted separately for faba bean and tomato yields in both sole and intercropped conditions, following the methodology established by Gomez and Gomez (1984). Mean comparisons were performed using the Least Significant Difference (LSD) test at the 5% significance level, as specified by Waller and Duncan (1969).

Findings and Elaborations

Weather Conditions at the Experimental Site

Figure 5.2a exhibited less variability in solar radiation over the two growing seasons. Elevated mean temperatures were seen at both the onset and conclusion of the second growing season, unlike the first growing season. Additionally, a comparable pattern in the variations of relative humidity was noted in both October and November, with elevated values recorded in the initial growing season compared to subsequent growth seasons (Fig. 5.2b). In contrast, substantial variations in wind speed were seen during the first growing season, while few variations were recorded in the second growth season (Fig. 5.2b). Consequently, the ETo values (Fig. 5.1c) were rather high at the beginning and end of the first growing season, while remaining lower for the majority of the season. During the second growing season, the ETo

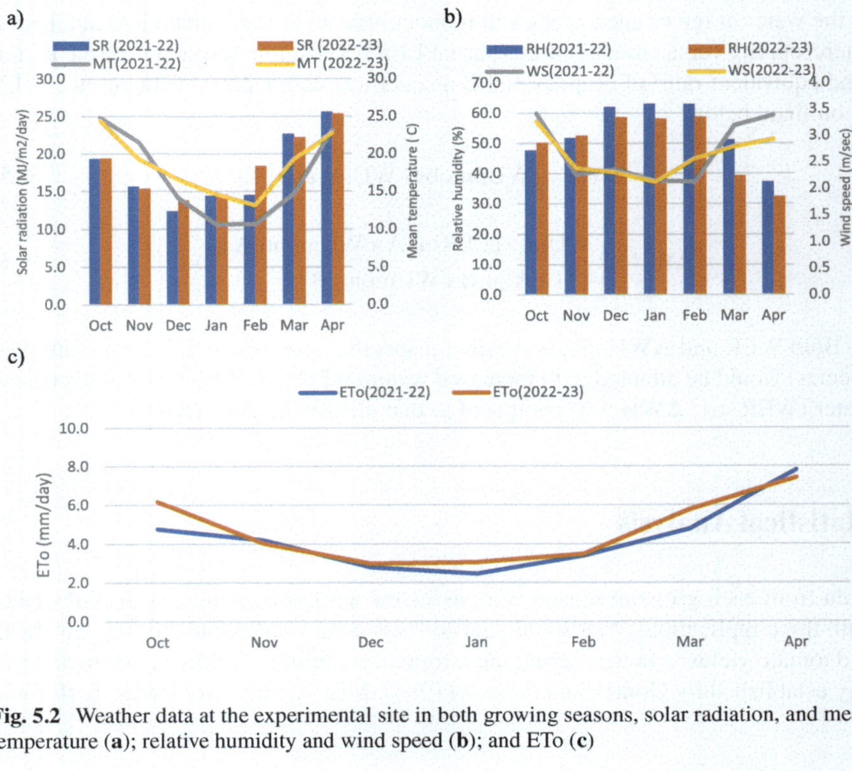

Fig. 5.2 Weather data at the experimental site in both growing seasons, solar radiation, and mean temperature (**a**); relative humidity and wind speed (**b**); and ETo (**c**)

Table 5.3 Applied irrigation water (mm) to sole tomato, sole faba bean and their intercropping systems in both growing seasons

	First growing season			Second growing season		
	RI	WS_1	WS_2	RI	WS_1	WS_2
Sole tomato	393	334	275	431	366	301
Sole faba bean	346	294	242	381	323	266
25% faba bean+100% tomato	393	334	275	431	366	301
50% faba bean+100% tomato	393	334	275	431	366	301
75% faba bean+100% tomato	393	334	275	431	366	301

RI, WS_1, and WS_2 = irrigation with 100%, 80%, and 75% ETc, respectively

values were significantly elevated compared to those in the first growth season (Fig. 5.1c). The value of the required irrigation water and water consumption for the studied cropping systems is expected to be higher in the second season compared to the first season.

The Irrigation Volumes and the Water Consumption of both Sole and Intercropped Crops

Table 5.3 outlines the projected volumes of irrigation water employed in the treatments for both sole tomato and its intercropping systems with faba bean, as well as for sole faba bean. The irrigation volumes administered to the sole tomato were equivalent to those utilized in the intercropping systems. The table's results indicated an increase in the applied irrigation water during the second growing season, relative to the first, due to elevated ETo values (Fig. 5.2c). Water stress treatments of 85% and 70% ETc resulted in a 16% and 31% decrease in irrigation water administered to sole crops and intercropping systems, respectively, averaged over both growing seasons, compared to the required irrigation quantities. Khalifa (2023) achieved analogous results by implementing water stress treatments at 85% and 70% ETc, resulting in a 15% and 30% reduction in applied water compared to the control value. He also noticed an increase in the irrigation water applied during the second growing season compared to the first season. Moreover, faba bean and tomato intercropping systems utilized the irrigation water given to the tomato, hence enhancing the efficient and complementary utilization of the irrigation water, as noted by Qin et al. (2013).

Table 5.4 demonstrated that the water consumption estimates for tomato intercropping systems exceeded those of sole tomato, due to the incorporation of intercropped faba bean at different planting densities. This result indicated that water loss through evapotranspiration and/or deep percolation diminished as the ground cover of intercropped faba bean increased, with the most pronounced reduction in water loss observed at 75% faba bean intercropped with 100% tomato, and the least reduction at 25% faba bean intercropped with 100% tomato. Moreover, elevated

Table 5.4 Water consumptive use (mm) of sole tomato, sole faba bean, and their intercropping systems

	First growing season			Second growing season		
	RI	WS_1	WS_2	RI	WS_1	WS_2
Sole tomato	342	287	237	375	315	253
Sole faba bean	325	276	218	331	275	226
25% faba bean+100% tomato	354	304	245	354	301	238
50% faba bean+100% tomato	365	311	256	400	340	280
75% faba bean+100% tomato	373	317	261	418	355	292

RI, WS_1, and WS_2 = irrigation with 100%, 85%, and 70% ETc, respectively

water consumption was seen in the second growing season relative to the first, along with increasing monthly ETo values (Fig. 5.2c). Reports indicate that intercropping may result in higher total water consumption relative to sole cropping; yet, it exhibits its superior water-usage efficiency, particularly when accounting for enhanced yield (Wang et al., 2015). Soil evaporation and soil water content were diminished in intercropping systems relative to sole planting, indicating more efficient water utilization (Rahman et al., 2017). Moreover, intercropping can alter the microclimate by decreasing daytime relative humidity and enhancing night-time relative humidity, thus minimizing plant water loss during the day and aiding in moisture retention at night (He et al., 2011).

The Yields of Tomato under Various Irrigation Regimes and Cropping Systems

The yield of tomato was markedly affected by the irrigation regime, cropping systems, and their interaction throughout both growing seasons ($p \leq 0.05$; Table 5.5). Under the three irrigation regimes, intercropping faba bean with tomato yielded a higher tomato production than its sole cultivation. Moreover, augmenting faba bean planting density enhanced tomato yield throughout the three intercropping systems, with the peak yield observed with the provision of required irrigation (RI). The intercropped tomato production increased by 42%, 36%, and 30% when 75% of faba bean was intercropped with tomato and irrigated with RI, WS1, and WS2, respectively, compared to the yield of sole tomato averaged over the two growing seasons. Ibrahim et al. (2010) reported analogous findings, indicating that intercropping faba bean with tomato enhanced the production by 36% relative to tomato cultivated in monoculture. Furthermore, the table showed that tomato yield was higher in the second growing season, relative to the first growing season.

The season length of sole tomato plants was approximately 45 days shorter than that of intercropped tomato, which adversely affected tomato yield. The extension

Table 5.5 The tomato yields (ton/ha) under different cropping systems (CS) and irrigation treatments (IR) in both growing seasons

Cropping system	2021/22 growing season				2022/23 growing season			
	RI	WS_1	WS_2	Mean	RI	WS_1	WS_2	Mean
Sole tomato	56.11	50.55	47.62	51.43	56.98	50.53	47.21	51.57
25% faba bean+100% tomato	65.05	52.12	50.73	66.24	53.47	50.98	49.73	50.12
50% faba bean+100% tomato	67.61	57.22	52.98	59.27	68.81	58.42	54.58	60.60
75% faba bean+100% tomato	79.79	67.89	61.39	69.69	80.99	69.09	62.32	70.80
Mean	67.14	55.82	51.43	58.13	68.25	56.63	52.27	59.05
$LSD_{0.05}$								
IR	1.89				1.45			
CS	0.94				1.38			
IR X CS	1.62				2.39			

RI, WS_1, and WS_2 = irrigation with 100%, 85%, and 70% ETc, respectively

of tomato-growing season is due to the intercropping of faba bean, which modulates the microclimate of tomato plants and alleviate cold stress. Moreover, as a leguminous crop, faba bean enhances nitrogen fixation (Kirihetti, 2018), increases available phosphorus (Mat Hassan et al., 2012), and augments soil organic matter (Banik et al., 2018), thus improving soil fertility through the residual effects of legumes, particularly in the subsequent growing season. An enhancement in soil nutrient availability and tomato yield was documented by intercropping of faba bean–tomato (Warren Raffa et al., 2022).

The maximum tomato yields were consistently achieved by the use of RI and intercropping 75% faba bean with tomato, specifically 79.79 ton/ha and 80.99 ton/ha in the first and second growing seasons, respectively. An adequate water supply under required irrigation likely facilitated optimal physiological processes, including photosynthesis, nutrient absorption, leading to enhanced tomato yield. Conversely, diminishing irrigation levels resulted in a gradual decrease in production. Moderate water stress (WS1) imposed on 75% faba bean intercropped with 100% tomato diminished tomato yield by roughly 15%, but severe water stress (WS2) led to yield losses of nearly 23% on average across both growth seasons, compared to the values obtained under RI. The results indicated a significant reliance of tomato productivity on the availability of irrigation water, as observed by Khalifa (2023). The results also showed that irrigation with WS1 and WS2 and intercropping 75% faba bean with tomato resulted in tomato yield higher than that obtained under sole tomato planting in both growing seasons.

The above results clearly demonstrate that intercropping tomato with faba bean substantially modifies the microclimate surrounding tomato plants, thereby improving crop performance under both optimal and stress conditions. Intercropping also influenced soil microclimate by enhancing ground cover, thereby reducing soil evaporation and maintaining more stable soil moisture conditions. This was reflected in the higher water consumption efficiency observed in intercropping systems relative to sole tomato cropping. The structural characteristics of faba bean, particularly its relatively taller canopy and rapid early-season growth, play a critical role in shaping the microenvironment within the intercropped canopy. These structural effects influence key microclimatic variables, including temperature, wind speed, radiation interception, and humidity, which collectively determine tomato physiological performance and productivity.

The Yield of Sole and Intercropped Faba Bean under Different Irrigation Treatments

Table 5.6 reveals significant differences ($p \leq 0.05$) between the sole and intercropped faba bean yields under the three irrigation treatments and the interaction between irrigation treatments and cropping systems in both seasons. The findings indicated an enhancement in the production of intercropped faba beans with an increase in planting density from 25% to 75% of the recommended rate over both growing seasons and under the three irrigation treatments. Moreover, the yield of both sole and intercropped faba bean was enhanced in the second growing season in comparison to the first season.

The implementation of RI achieved the maximum yield of both sole and intercropped faba beans. Application of water stress treatments diminished the yield of sole faba beans by 19% and 21% under WS1 and WS2, respectively, averaged over both growing seasons, relative to the yields obtained under RI. The production losses for intercropped faba bean at 75% of its recommended rate with tomato were 25% and 39% under irrigation at WS1 and WS2, respectively, averaged over two seasons, in contrast to the values obtained under RI (Table 5.6).

Table 5.6 The faba bean yields (ton/ha) under different cropping systems (CS) and irrigation treatments (IR) in both growing seasons

Cropping system	2021/22 growing season				2022/23 growing season			
	RI	WS_1	WS_2	Mean	RI	WS_1	WS_2	Mean
Sole faba bean	4.95	4.01	3.89	4.28	5.07	4.11	3.99	4.39
25% faba bean+100% tomato	1.36	1.09	1.02	1.16	1.57	1.18	1.12	1.29
50% faba bean+100% tomato	2.16	1.64	1.55	1.79	2.97	1.93	1.62	2.17
75% faba bean+100% tomato	2.78	2.05	1.82	2.22	3.18	2.44	1.81	2.48
Mean	2.81	2.20	2.07	2.36	3.20	2.41	2.13	2.58
$LSD_{0.05}$								
IR	0.23				0.32			
CS	0.25				0.22			
IR X CS	0.43				0.38			

RI, WS_1, and WS_2 = irrigation with 100%, 85%, and 70% ETc, respectively

Khan et al. (2010) reported that faba bean has more vulnerability to both terminal and transitory drought than other temperate grain legumes, indicating it may have competed more intensely with tomato for water during conditions of water stress.

Table 5.6 indicates that the output of intercropped faba bean increased by 25% of its planting density compared to its intercropped percentage, achieving an average yield of 29% throughout the three irrigation regimes. Intercropped faba bean has been shown to enhance root intensity and depth relative to sole cropping, potentially benefiting yield (Shanmugam et al., 2022). Conversely, the yields of intercropped faba beans at 50% and 75% of their rate were inferior to their respective percentages. This may be attributed to the greater intraspecific competition among faba bean plants compared to the interspecific competition between faba bean and tomato plants. López-Bellido et al. (2005) assert that faba bean output rises with plant density until an ideal threshold is reached, after which additional increases in density do not augment yield and may even diminish seed production due to heightened competition among plants.

Land Equivalent Ratio (LER)

The values of LER_{bean}, LER_{tomato}, and LER_{total} in Table 5.7 revealed significant differences ($p \leq 0.05$) for irrigation treatments, cropping systems, and the interaction between them in both growing seasons.

The findings indicated that augmenting faba bean planting density enhanced LER_{bean}, LER_{tomato}, and LER_{total} across the three irrigation treatments in both growing seasons. In both growth seasons, the LER_{bean} values were below 1.0 across all three irrigation treatments, attributed to a planting density less than 100%. The highest value recorded was 0.57 for faba bean intercropped at 75% and irrigated with RI. The values of LER_{tomato} were found to exceed 1.0 across the three intercropping systems irrigated with RI, attributed to the elevated intercropped tomato yield, with the peak value reaching 1.42 when 75% faba bean was intercropped with tomato, averaged over both growth seasons (Table 5.7).

In water stress treatments, intercropping 25% faba bean resulted in LER_{tomato} values below 1.0; however, increasing faba bean density to 50% and 75% elevated the LER_{tomato} to exceed 1.0. The results also indicated that intercropping with faba bean at planting densities of 50% to 75% can alleviate the impact of water stress on tomato productivity (Table 5.7).

The maximum LER_{total} values, specifically 2.02, were achieved when 75% faba bean intercropped with tomato under irrigation with RI, averaged across both growth seasons. These results implied that if faba bean had been cultivated as a monocrop, 102% more land would have been required to attain the equivalent

Table 5.7 The land equivalent ratios (LER) as affected by tomato intercropping systems (IS) under the studied irrigation treatments (IR) in both growing seasons

		2021/22 growing season			2022/23 growing season		
IR	IS	LER_{bean}	LER_{tomato}	LER_{total}	LER_{bean}	LER_{tomato}	LER_{total}
RI	25% faba bean+100% tomato	0.28	1.16	1.43	0.31	1.16	1.47
	50% faba bean+100% tomato	0.44	1.20	1.65	0.59	1.21	1.80
	75% faba bean+100% tomato	0.57	1.41	2.00	0.63	1.42	2.05
	Mean	0.43	1.26	1.69	0.51	1.26	1.78
WS_1	25% faba bean+100% tomato	0.27	0.94	1.21	0.29	0.96	1.25
	50% faba bean+100% tomato	0.41	1.13	1.54	0.47	1.16	1.63
	75% faba bean+100% tomato	0.51	1.34	1.85	0.59	1.37	1.96
	Mean	0.40	1.14	1.54	0.45	1.16	1.61
WS_2	25% faba bean+100% tomato	0.26	0.92	1.18	0.28	0.95	1.23
	50% faba bean+100% tomato	0.40	1.11	1.51	0.41	1.16	1.56
	75% faba bean+100% tomato	0.47	1.29	1.76	0.45	1.32	1.77
	Mean	0.38	1.11	1.48	0.38	1.14	1.52
Mean (IS)	25% faba bean+100% tomato	0.27	1.01	1.28	0.29	1.26	1.32
	50% faba bean+100% tomato	0.42	1.15	1.57	0.49	1.16	1.66
	75% faba bean+100% tomato	0.52	1.35	1.87	0.56	1.14	1.93
Mean		0.40	1.17	1.57	0.45	1.19	1.64
$LSD_{0.05}$							
IR		0.10	0.06	0.11	0.09	0.01	0.01
IS		0.04	0.02	0.05	0.04	0.03	0.05
IR X IS		0.08	0.04	0.09	0.07	0.05	0.08

RI, WS_1, and WS_2 = irrigation with 100%, 85%, and 70% ETc, respectively

tomato output of the intercropping system. Upon the implementation of water stress treatments, the total LER values diminished, namely to 1.91 and 1.77 for WS1 and WS2, respectively, averaged across both growing seasons. Abd El-Aal and Zohry (2004) and Ibrahim et al. (2010) produced analogous results.

The results showed that, at all planting densities, the intercropping system exhibited strong evidence of resource complementarity between faba bean and tomato. Complementarity was evident in the efficient use of light, water, and soil nutrients, resulting in increased tomato yield and land-use efficiency relative to monocropping. The partial land equivalent ratio of tomato consistently exceeded unity under these densities, indicating that tomato benefited from the presence of faba bean rather than experiencing yield suppression.

Although increasing faba bean density strengthened microclimatic modification and enhanced tomato yield, it also intensified competition for resources, particularly under water stress. At higher densities (75% of the recommended rate), the system approached a threshold where facilitative effects and competitive interactions coexisted. The highest tomato yields and system productivity were achieved at this density, indicating that microclimatic and complementary effects still outweighed competitive pressures under adequate irrigation.

Water Equivalent Ratio (WER) and Change in Water Use (ΔWU)

Table 5.8 indicates significant variations ($p \leq 0.05$) in WER_{bean}, WER_{tomato}, and WER_{total} among irrigation regimes, intercropping system and the interaction between them. Augmenting faba bean planting density elevated the values of WER_{bean}, WER_{tomato}, and WER_{total} in both growing seasons. The WER_{bean} values were determined to be below 1.0 across various irrigation treatments and intercropping systems. Increasing faba bean planting density elevated the values of WER_{bean} throughout the three irrigation treatments, with the highest values seen under the application of RI, namely 0.50. The results indicated that the WER_{tomato} values were above 1.0 across all three irrigation treatments and intercropping systems, with the maximum value of 1.29 recorded when 75% faba bean was intercropped with tomato under RI application. The WER_{total} values surpassed 1.0, particularly when 75% faba bean was intercropped with tomato under RI irrigation, reaching 1.80. This figure demonstrates that the intercropping system may attain a designated yield with diminished water use compared to achieving the same yield in separate monoculture fields, suggesting an 80% improvement in water-use efficiency at the system level in contrast to monoculture.

Under water stress treatments, the values of WER_{total} decreased to 1.76 and 1.63 with irrigation using WS1 and WS2, respectively, averaged over two growing seasons, in contrast to the values recorded under required irrigation (RI). Zhang et al. (2022) assert that WER markedly improves water usage efficiency compared to monocropping systems by facilitating complementary water absorption at various soil depths and growth phases, therefore minimizing water losses and enhancing total water production.

Marked changes ($p \leq 0.05$) were noted in ΔWU values among irrigation treatments, tomato intercropping systems, and their interaction during both growing seasons (Table 5.8). The findings indicated that irrigation under RI conditions diminished ΔWU values in comparison to irrigation under water stress. Consistently

Table 5.8 Water equivalent ratio (WER) and change in water use (ΔWU) under irrigation treatments and tomato intercropping systems in both growing seasons

		2021/22 growing season			ΔWU	2022/23 growing season			ΔWU
IR	IS	WER_{bean}	WER_{tomato}	WER_{total}	(%)	WER_{bean}	WER_{tomato}	WER_{total}	(%)
RI	IS1	0.25	1.12	1.37	−0.27	0.26	1.12	1.39	−0.28
	IS2	0.39	1.13	1.52	−0.34	0.48	1.13	1.61	−0.38
	IS3	0.49	1.30	1.79	−0.44	0.5	1.31	1.81	−0.44
	Mean	0.38	1.18	1.56	−0.35	0.42	1.18	1.59	−0.37
WS_1	IS1	0.25	0.89	1.21	−0.12	0.24	0.91	1.14	−0.13
	IS2	0.36	1.05	1.54	−0.29	0.38	1.07	1.45	−0.31
	IS3	0.44	1.22	1.85	−0.40	0.46	1.21	1.67	−0.40
	Mean	0.35	1.05	1.54	−0.27	0.36	1.06	1.42	−0.28
WS_2	IS1	0.24	0.89	1.18	−0.11	0.23	0.88	1.11	−0.10
	IS2	0.35	1.03	1.51	−0.27	0.33	1.04	1.37	−0.27
	IS3	0.4	1.17	1.76	−0.36	0.35	1.14	1.49	−0.33
	Mean	0.38	1.11	1.48	−0.25	0.38	1.14	1.52	−0.23
Mean (IS)	IS_1	0.27	1.01	1.28	−0.17	0.29	1.26	1.32	−0.17
	IS_2	0.42	1.15	1.57	−0.30	0.49	1.16	1.66	−0.32
	IS_3	0.52	1.35	1.87	−0.40	0.56	1.14	1.93	−0.39
	Mean	0.40	1.17	1.57	−0.29	0.45	1.19	1.64	−0.29
$LSD_{0.05}$									
IR		0.03	0.01	0.02	0.03	0.01	0.02	0.03	0.05
IS		0.01	0.01	0.02	0.02	0.02	0.01	0.02	0.01
IR × IS		0.02	0.02	0.03	0.03	0.03	0.01	0.03	0.02

IR = irrigation systems; IS = intercropping systems; RI, WS_1, and WS_2 = irrigation with 100%, 85%, 70% ETc, respectively; IS_1 = 25% faba bean+100% tomato; IS_2 = 50% faba bean+100% tomato; IS_3 = 75% faba bean+100% tomato

lower ΔWU values were observed under RI, signifying a more pronounced water-saving benefit compared to the water-stress treatments. This tendency was seen across irrigation regimes, intercropping systems, and their interaction, indicating that the intercropping system attained superior water-use efficiency compared to sole cropping, especially under sufficient water supply. Conversely, sole crops may have utilized significantly more water than intercrops under required irrigation, thus expanding the water-saving disparity (Mao et al., 2012). Moreover, all assessed intercropping systems demonstrated WER_{total} values above 1.0 and ΔWU values below 0, hence affirming improved water-use efficiency (Table 5.8). Similar WER and ΔWU responses have been documented in Egyptian agroecosystems, encompassing sugar beet–legumes intercropping (Ouda & Zohry, 2023), wheat–legumes intercropping systems (Zohry & Ouda, 2025), and soybean–maize intercropping systems (Ouda & Zohry, 2026).

The results showed that water-use indicators, namely WER_{total} and ΔWU, further support the complementarity hypothesis. Water equivalent ratio values exceeding unity and negative ΔWU values across intercropping treatments indicate that the system achieved higher yields per unit of water than monocropping. This reflects complementary water uptake patterns and improved water-use efficiency at the system level, particularly under sufficient irrigation. Nevertheless, even under conditions where competition intensified under water stress, the net system performance remained superior to monocropping, demonstrating that facilitative interactions—particularly microclimate modification and improved resource-use efficiency—continued to play a dominant role. This dynamic balance between complementarity and competition suggests that optimal faba bean density should maximize facilitative interactions while minimizing competitive pressures, with 50–75% of the recommended density representing an effective range under Egyptian conditions.

Conclusion

This study illustrates that intercropping faba bean with tomato is an effective technique for enhancing tomato productivity, resource-use efficiency, and climate resistance under Egyptian agroecological conditions. The intercropping technique significantly altered the microclimate around tomato plants by diminishing cold stress exposure, stabilizing temperature variations, and prolonging the growing season for tomatoes. The microclimatic enhancements substantially improved fruit set, elevated yields, and augmented marketable production relative to monoculture tomato farming.

Augmenting faba bean planting density enhanced system performance, with the 75% faba bean density consistently attaining the maximum tomato output, land-use efficiency, and water-use efficiency across irrigation treatments. The intercropping system exhibited distinct benefits for land equivalent ratio and water equivalent ratio, both surpassing unity across treatments, thereby validating a more efficient utilization of land and water resources compared to monocropping systems.

Water stress markedly decreased the productivity of both tomato and faba bean; nonetheless, intercropping alleviated these adverse effects by enhancing water-use efficiency and sustaining comparatively higher yields under deficit irrigation. The integrated assessment of WER and ΔWU further validated the preeminence of intercropping systems in attaining greater production per unit of water utilized. The findings underscore the potential of incorporating legumes, such as faba bean, into tomato production systems to improve resistance to water scarcity and climate unpredictability.

Faba bean–tomato intercropping constitutes a viable strategy for sustainable intensification in Egypt and comparable arid and semi-arid areas. The implementation of suitable planting densities and irrigation management can optimize microclimate regulation, promote resource-use efficiency, and improve agricultural output. Subsequent research should investigate the long-term impacts on soil fertility, the dynamics of pests and diseases, and the economic viability to facilitate broader implementation of this climate-smart intercropping system.

References

Abd El-Aal, A. I. N., & Zohry, A. A. (2004). Land and water exploitation for maximizing productivity of tomato intercropped with faba bean crop under stress ecological conditions at Toshky. *Egyptian Journal of Agricultural Research, 82*(1), 37–45.

Abd El-Hady, M. A., Marey, R. A., & Abou-Keriasha, M. A. (2013). Protection of tomato plants against cold and frost conditions by intercropping under Sohag governorate. *Journal of Plant Production, 4*(7), 1077–1096.

Abdelhaleim, M. S., El-Ghobashy, A. A., El-Sayed, S. M., Abdelmageed, A. A. a., & El-Ashry, S. M. (2022). Assessment of drought tolerance indices in faba bean (Vicia faba L.) genotypes under varied irrigation regimes. *Plants, 11*(2), 300. https://doi.org/10.3390/plants11020300. (Note: This is cited as a "2022 study" — adjust if you have the exact article details.).

Abid, G., Chairi, F., Hazzoumi, Z., Bakhat, H., Nabloussi, A., el-Fadli, M., & El Modafar, C. (2017). Agro-physiological and biochemical responses of faba bean (*Vicia faba* L.) to water deficit stress. *Journal of Agronomy and Crop Science, 203*(4), 361–372. https://doi.org/10.1111/jac.12261

Ali, E. M. A. (2026). An economic study of the factors affecting tomato exports in Egypt. *Alexandria Journal of Agricultural Sciences, 71*(1), 24–45. https://doi.org/10.21608/alexja.2025.413831.1162

Allen, R. G., Pereira, L. S., Raes, D., & Smith, M. (1998). *Crop evapotranspiration: Guideline for computing crop water requirements* (pp. 109–158). FAO. No56.

Alshameri, A. M., Alghamdi, S. S., Gaafar, A. Z., Almunqedhi, B. M., Qahtan, A. A., Hodhod, M. S. a., & Migdadi, H. M. (2021). Effect of water deficit on yield of different faba bean (Vicia faba L.) genotypes. *Legume Research, 44*(6), 718–722. https://doi.org/10.18805/LR-600

Banik, D., Upadhyay, H., Devi, B. P., & Darvhankar, M. (2018). Effect of different irrigation system on growth and development of wheat-chickpea intercropping. *Bulletin of Environmental Pharmacology and Life Sciences, 8*(1), 32–34.

Daryanto, S., Wang, L. a., & Jacinthe, P. A. (2015). Global synthesis of drought effects on cereal, legume, and oilseed crops production: A review. *Agricultural Water Management, 147*, 30–41. https://doi.org/10.1016/j.agwat.2014.05.021

El-Mehy, A. A., & Mohamed, M. H. M. (2018). Yield and economic evaluation of maize and tomato as affected by cropping systems and some growth stimulants. *Middle East Journal of Applied Sciences, 8*(1), 209–222.

Elsadany, M. F. I., & El-Shamy, M. A. (2016). Influence of faba bean-tomato intercropping on mites, insects' infestation and the relationship between the pests and predators, leaf phenols content, and yield components of both crops in different intercropping systems. *Mansoura Journal of Biology*, 55–68.

Gomez, K. A., & Gomez, A. A. (1984). *Statistical procedures for agricultural research* (2nd ed., pp. 22–67). John Wiley and Sons.

Hajihashemi, S. (2019). Stomatal regulation as a drought-tolerance mechanism. In A. Roychoudhury & D. Tripathi (Eds.), *Molecular plant abiotic stress*. https://doi.org/10.1002/9781119463665.ch3

Hatamleh, A. A., Danish, M., Munirah, B., Al-Dosary, A., El-Zaidy, M., & Ali, S. (2022). Physiological and oxidative stress responses of *Solanum Lycopersicum* (L.) (tomato) when exposed to different chemical pesticides. *RSC Advances, 12*, 7237–7252.

He, H., Yang, L., Fan, L., Zhao, L., Wu, H., et al. (2011). The effect of intercropping of maize and soybean on microclimate. In *5th computer and computing technologies in agriculture (CCTA), Beijing, China* (pp. 257–263) ff10.1007/978-3-642-27278-3_27ff.fhal-01360987f.

Huang, C., Cheng, P., Zhang, H., Jiang, C., & Jin, L. (2022). Identification and differential expression of cold-stress-responsive microRNAs in cold-tolerant and -susceptible hemerocallis fulva varieties. *New Zealand Journal of Crop and Horticultural Science, 50*, 1–15. https://doi.org/10.1080/01140671.2021.2013260

Ibrahim, S., Shaaban, M., & Gendy, E. K. (2010). Intercropping faba bean with tomato. *Egyptian Journal of Applied Science, 25*(6A), 167–181.

Khalifa, R. M. (2023). Effects of inadequate drip irrigation and organo-mineral fertilizer on tomato production, quality and water use efficiency in sandy soils. *Egyptian Journal of Soil Science, 63*(3), 325–338.

Khan, H. R., Paull, J. G., Siddique, K. H. M., & Stoddard, F. L. (2010). Faba bean breeding for drought-affected environments: A physiological and agronomic perspective. *Field Crops Research, 115*(3), 279–286. https://doi.org/10.1016/j.fcr.2009.09.003

Kirihetti, D. (2018). *Direct assessment of the fixation and release of nitrogen by pea, lentil, chickpea and Faba Bean. A master of science thesis*. University of Saskatchewan Saskatoon, Canada.

Klute, A. (1986). *Methods of soil analysis, Part I. Physical and mineralogical method*. American Society of Agronomy.

Liang, G., Liu, J., & Zhang, J. (2020). Effects of drought stress on photosynthetic and physiological parameters of tomato. *Journal of the American Society for Horticultural Science, 145*(1), 12–17. https://doi.org/10.21273/JASHS04725-19

Liu, H., Ouyang, B., Zhang, J., Wang, T., Li, H., Zhang, Y., et al. (2012). Differential modulation of photosynthesis, signaling, and transcriptional regulation between tolerant and sensitive tomato genotypes under cold stress. *PLoS One, 7*, e50785.

López-Bellido, F. J., López-Bellido, L., & López-Bellido, R. J. (2005). Competition, growth and yield of faba bean (Vicia faba L.). *European Journal of Agronomy, 23*(4), 359–378. https://doi.org/10.1016/j.eja.2005.02.002

Mao, L. L., Zhang, L. Z., Li, W. W., Werf, W. V. D., Sun, J. H., Spiertz, H., & Li, L. (2012). Yield advantage and water saving in maize/pea intercrop. *Field Crops Research, 138*, 11–20.

Masoued, S. R. (2017). *Climate and its impact on diseases in the northern upper Egypt (applied climatology study)*. Msc Thesis (pp. 15–36). Department of Geography, Cairo University, Egypt.

Mat Hassan, H., Marschner, P., McNeill, A., & Tang, C. (2012). Grain legume pre-crops and their residues affect the growth, P uptake and size of P pools in the rhizosphere of the following wheat. *Biology and Fertility of Soils, 48*, 775–785. https://doi.org/10.1007/s00374-012-0671-8

Meena, Y., Khurana, D., Kaur, N., & Singh, K. (2018). Towards enhanced low temperature stress tolerance in tomato: An approach. *Journal of Environmental Biology, 39*, 529–535.

Mesa, T., Polo, J., Arabia, A., Caselles, V., & Munné-Bosch, S. (2022). Differential physiological response to heat and cold stress of tomato plants and its implication on fruit quality. *Journal of Plant Physiology, 268*, 153581. https://doi.org/10.1016/j.jplph.2021.153581

Ministry of Agriculture and Land Reclamation. (2018). Agricultural Statistical Bulletin. *Part one: Winter crops*. Economic Affairs Sector. Egypt.

Moreira de Carvalho, D. B., de Souza, A. V., Domingues Neto, F. J., Iglesias, L., Nogueira, B. B., Zanetti, W. A., Gabriel Filho, L. R. A., & Putti, F. F. (2020). Growth models and Nutriente uptake curve of tomato BRS 1509 subjected to irrigation pulses. *Journal of Plant Nutrition, 43*, 1999–2012.

Morris, R. A., & Garrity, D. P. (1993). Resource capture and utilization in intercropping: Water. *Field Crops Research, 34*, 303–317.

Muktadir, M. A., Ali, M. A., Khan, H., Siddique, K. H. M. a., & Lam-Shoemaker, H. (2020). Physiological and biochemical basis of faba bean (Vicia faba L.) responses to drought stress: Implications for breeding. *Agronomy, 10*(9), 1345. https://doi.org/10.3390/agronomy10091345

Ouda, S., & Zohry, A. (2023). Increasing sustainability by intercropping legume crops with sugar beet under imposed water stress. *Irrigation and Drainage*, 1–26. https://doi.org/10.1002/ird.2884

Ouda, S. a., & Zohry, A. (2026). Optimizing resource utilization through intercropping soybeans (*Glycine max* (L.) Merr.) and maize (*Zea mays* L.) in Middle Egypt. *Acta Agriculturae Slovenica, 122*(1), 1–23.

Petrović, I., Jovanović, Z., Stikić, R., Marjanović, M. a., & Savić, S. (2021). Effect of long-term moderate drought on metabolic and antioxidative responses of tomato leaves. *Physiology and Molecular Biology of Plants, 27*(1), e01102. https://doi.org/10.1007/s12298-021-01102-2

Putti, F. F., de Queiroz Barcelos, J. P., Goes, B. C., Alves, R. F., Neto, M. M., da Silva, A. O., Filho, L. R. A. G., Zanetti, W. A. L., & de Souza, A. V. (2023). Effects of water deficit on growth and productivity in tomato crops irrigated with water treated with very low-frequency electromagnetic resonance fields. *Plants, 12*, 3721. https://doi.org/10.3390/plants12213721

Qin, A. Z., Huang, G. B., Chai, Q., Yu, A. Z., & Huang, P. (2013). Grain yield and soil respiratory response to intercropping systems on arid land. *Field Crops Research, 144*, 1–10. https://doi.org/10.1016/j.fcr.2012.12.005

Rahman, T., Liu, X., Hussain, S., Ahmed, S., & Chen, G. Y. F. (2017). Water use efficiency and evapotranspiration in maize-soybean relay strip intercrop systems as affected by planting geometries. *PLoS One, 12*(6), e0178332. https://doi.org/10.1371/journal.pone.0178332

Raja, V., Qadir, S. U., Alyemeni, M. N., & Ahmad, P. (2020). Impact of drought and heat stress individually and in combination on physio-biochemical parameters, antioxidant responses, and gene expression in *Solanum Lycopersicum. 3 Biotech, 10*, 208.

Ramadan, A. A., El-Bassiouny, H. M. S. a., & El-Karamany, M. F. (2025). Enhancing faba bean (*Vicia faba* L.) productivity under drought stress through modulation of physiological traits and antioxidant enzyme system using thiourea and hydrogel. *Plants, **14***, 40615955. https://doi.org/10.3389/fpls.2025.12231906

Rao, M. R., & Willey, R. W. (1980). Evaluation of yield stability in intercropping: Studies on sorghum/pigeon pea. *Experimental Agriculture, 16*(2), 105–116.

Shanmugam, S., Hefner, M., Pelck, J. S., Labouriau, R., & Kristensen, H. L. (2022). Complementary resource use in intercropped faba bean and cabbage by increased root growth and nitrogen use in organic production. *Soil Use and Management, 38*, 729–740. https://doi.org/10.1111/sum.12765

Sheha, A. M., El-Mehy, A. A., Mohamed, A. S., et al. (2022). Different wheat intercropping systems with tomato to alleviate chilling stress, increase yield and profitability. *Annals of Agricultural Sciences, 67*, 136–145.

Taha, A., & Kasem, E. (2022). Water management of maize-cowpea intercropping system under surface irrigation. *Moroccan Journal of Agricultural Sciences, 3*(1), 29–38.

Tan, K. H. (1996). *Soil sampling, preparation and analysis* (p. 391). Marcel Dekker, Inc..

Tanny, J. (2013). Microclimate and evapotranspiration of crops covered by agricultural screens: A review. *Biosystems Engineering, 114*(1), 26–43. https://doi.org/10.1016/j.biosystemseng.2012.10.008

Vermeiren, L., & Jopling, G. A. (1984). *Localized irrigation.* FAO irrigation and drainage paper no. 36, Rome, Italy.

Wadood, A., et al. (2024). Unraveling the impact of water deficit stress on nutritional and defense responses in tomato. *Frontiers in Plant Science, **15***, 1403895. https://doi.org/10.3389/fpls.2024.1403895

Waller, R. A., & Duncan, D. B. (1969). *Bays rule for the symmetric multiple comparison problem.* American State Association.

Wang, Z., Wu, P., Zhao, X., Gao, Y., & Chen, X. (2015). Water use and crop coefficient of the wheat–maize strip intercropping system for an arid region in northwestern China. *Agricultural Water Management, 161*, 77–85. https://doi.org/10.1016/j.agwat.2015.07.012

Wang, L.-h., Tan, D.-h., Zhong, X.-s., Jia, M.-q., Ke, X., Zhang, Y.-m., Cui, T., & Shi, L. (2024). Review on toxicology and activity of tomato glycoalkaloids in immature tomatoes. *Food Chemistry, 447*, 138937. https://doi.org/10.1016/j.foodchem.2024.138937

Warren Raffa, D., Migliore, M., Campanelli, G., Leteo, F., & Trinchera, A. (2022). Effects of Faba bean strip cropping in an outdoor organic tomato system on soil nutrient availability, production, and N budget under different fertilizations. *Agronomy, 12*, 1372. https://doi.org/10.3390/agronomy12061372

Yousef, E. A. A. a., & Ali, M. A. M. (2019). Alleviation of cold stress on tomato during winter season by application of yeast extract and glycinebetaine. *Egyptian Journal of Horticulture, 46*(1), 117–131. https://doi.org/10.21608/ejoh.2019.8020.1096

Zhang, Y., Sun, Z., Su, Z., Du, J., Bai, W., Wang, Q., Wang, R., et al. (2022). Root plasticity and interspecific complementarity improve yields and water use efficiency of maize/soybean intercropping in a water-limited condition. *Field Crops Research, 282*, 108523. https://doi.org/10.1016/j.fcr.2022.108523

Zohry, A., & Ouda, S. (2025). Legume crops enhance water use efficiency under intercropping system with wheat. *Moroccan Journal of Agricultural Science, 6*(3), 145–160.

Part II
Support Services of Legume

Chapter 6
Legume-Based Cropping Systems Enhance Soil Nutrients Circulation

Abstract Legume-based cropping systems play a pivotal role in enhancing agroecosystem services through improving soil fertility and nutrient circulation. This chapter examines the contribution of legumes to nutrient cycling and soil health within diversified cropping systems, emphasizing their ecological and agronomic importance for sustainable intensification. Legumes enhance soil nutrient circulation through multiple mechanisms, including biological nitrogen fixation, organic matter addition, deep nutrient recycling, and mobilization of macro- and micronutrients. Symbiotic nitrogen fixation by legumes enriches soil nitrogen reserves and reduces dependence on synthetic fertilizers, while crop residues and root exudates stimulate microbial activity and improve soil structure and nutrient availability. Intercropping and crop rotation with legumes further enhance nutrient cycling by increasing soil organic matter, improving phosphorus and potassium availability, and promoting beneficial microbial communities. Evidence from field studies demonstrates that legume inclusion in cropping systems significantly increases soil available nitrogen, phosphorus, potassium, and organic matter, thereby improving productivity and resource-use efficiency. The extent of nutrient enhancement depends on factors such as legume species, planting density, planting date, and water availability. Although water stress can reduce nitrogen fixation, intercropping with legumes often mitigates these adverse effects by improving root interactions and nutrient mobilization under stress conditions. Overall, legume-based cropping systems contribute to improved soil fertility, enhanced nutrient-use efficiency, and greater agroecosystem sustainability.

Keywords Nitrogen fixation · Deep nutrient recycling · Phosphorus · Micronutrients · Soil organic matter

A. E. Zohry, S. A. H. Ouda, *The Agroecosystem Services of Leguminous Crops*, https://doi.org/10.1007/978-3-032-32461-0_6

Introduction

Agroecosystem services are the benefits humans get from agricultural lands, categorized as provisioning, regulating, supporting, and cultural services (Liu et al., 2022). These services, vital for well-being, depend heavily on management, with sustainable practices enhancing them and poor management creating disservices like erosion and pollution, emphasizing the need for integrated approaches (Gairola et al., 2025). Among the supporting services, agroecosystem services enhance soil fertility and nutrients circulation (Liu et al., 2022). Nutrient circulation refers to the essential ecological process where essential elements (like carbon, nitrogen, phosphorus) move between living organisms (biotic) and the non-living environment (abiotic) in a continuous, repeating loop (Deemy et al., 2022). According to Rana et al. (2018), improving soil pore structure, soil aggregation and organic matter are fundamental to nutrient acquisition by crops, as well as water retention. The cycling of nutrition elements such as nitrogen and phosphorus between the soil, crops and microorganism, is basis for maintaining soil fertility and productivity (Zabala et al., 2021).

Legume-based cropping systems are considered as a sustainable management practice that can enhance the services provided by the agroecosystem. Legumes exhibit multifaceted effects on nutrient accumulation and enhancement of soil fertility (Jhariya et al., 2021). Legumes form symbiotic relationships with rhizobia in the soil, fix atmospheric nitrogen (Quinones et al., 2022), facilitate nutrient cycling, and improve soil water retention (Mrunalini et al., 2022). The ability of legumes to fix nitrogen in the soil, enhancing soil fertility and general plant health (Blanchart et al., 2016), resulting in healthier plants that are typically less vulnerable to pest damage.

Crop diversification with sustainable cropping systems enhances the agroecosystem stability by improving soil health and increasing resilience to climatic and biotic stresses, thereby promoting sustainability (Kumari et al., 2024). Legume-based cropping systems enhance the physical, chemical, and biological attributes of soil, including pH, water availability, and nutrient availability (Abbas et al., 2013). The legume-based intercropping systems and root exudates foster a positive rhizobacterial population, enhancing soil quality and the rhizosphere environment for improved resource acquisition (Zhao et al., 2022). Legume-based cropping enhances soil microorganisms and soil biological health which is determined by the soil microbial biomass carbon, microbial biomass nitrogen, and enzymatic activity (Borase D. et al., 2020). Moreover, the microbial population structure is further affected by the physicochemical properties of the soil (Song et al., 2023).

Integration of legumes into the cropping system is considered as a sustainable practice, where it can be done in different forms, namely intercropping with non-legume crops, and/or rotating with non-legume crops. Intercropping is an agricultural practice used to grow two or more different crops in alternate rows on a given plot of farmland at the same time during their growing season or in a growing period (Federer, 2012). According to Xu et al. (2020), intercropping has been considered as the "new green revolution" because of the increased land and water-use efficiency, attained from the complementary, compensation and cooperation between the intercrops in a way to achieve sustainable agricultural intensification. Intercropping systems may enhance nutrient uptake capability by augmenting the number of absorptive roots or the density of root hairs, hence optimizing the absorption of nutrients and water (Homulle et al., 2022). Shen et al. (2024) indicated that intercropping leguminous crops with tiger nut markedly enhanced soil organic content and alkali-hydrolyzable nitrogen, attributable to an increase in bacterial populations involved in soil nutrient cycling, including rhizobia, phosphate-solubilizing, and potassium-solubilizing bacteria (Zhang et al., 2018). Additionally, legumes intercropping and/or rotating with non-legume crops dramatically enhance nutrient cycling, primarily by adding nitrogen to the soil, improving the availability of other nutrients like phosphorus and potassium, and increasing soil organic matter (Walia et al., 2025). Consequently, legumes significantly enhance soil fertility and nutrient cycling through a combination of nitrogen fixation, improved soil structure, and increased soil organic matter (Akchaya et al., 2025).

This chapter will tackle the role of legume-based cropping systems in providing the crucial agroecosystem for enhancing soil nutrient circulation through biological nitrogen fixation, the addition of organic matter, and deep nutrient cycling.

Key Mechanisms of Nutrient Circulation by Legumes

Atmospheric Nitrogen Fixation

The most significant contribution of legumes is the conversion of inert atmospheric nitrogen (N_2) into plant-available ammonia (NH_3) through a symbiotic relationship with Rhizobia bacteria that live in root nodules residing in their roots. These bacteria produce NH_3 from the hydrogen obtained from the plant's carbohydrates and nitrogen from the air. The NH_3 then provides a supply of nitrogen to the plant to grow (Yuvaraj et al., 2020). Shortly after a legume seed germinates in the presence of Rhizobia microorganism in the soil, the bacteria penetrate the root hairs and cross into the root itself, where they multiply, inflicting a swelling of the root to the shape of pale pink nodules (Naab et al., 2009). This process adds substantial amounts of nitrogen to the soil, benefiting the legume crop and subsequent non-legume crops in a rotation (Akchaya et al., 2025). Bhowmik and Das (2018) assert that biological nitrogen fixation can be enhanced by inoculating host legumes with compatible rhizobia, resulting in residual nitrogen in the soil, contributing to organic matter, and serving as a cost-effective nutrient source for subsequent cropping seasons. Furthermore, about two-thirds of the nitrogen fixed through a legume crop becomes handy the subsequent growing season after a legume in a rotation (Deakin & Broughton, 2009).

It has been reported that intercropping with legumes enhances soil nitrogen through plant residues (Amossé et al., 2014). Naudin et al. (2010) assert that interspecific competition for soil nitrogen between wheat and peas is less significant than intraspecific competition among wheat plants in monoculture, with wheat yields frequently demonstrating superiority in intercropping relative to sole cropping. Furthermore, intercropping cowpeas with maize improved soil health by increasing nitrogen fixation and facilitating the absorption of soil nitrogen by cereals (Boukar et al., 2013).

Legume Crops Preceding Non-legume Crops

Legumes used as preceding crops provide substantial agronomic and ecological benefits to subsequent non-legume crops, often referred to as the "rotation effect." This advantage extends beyond nitrogen supply and includes improvements in soil structure and nutrient availability (Preissel et al., 2015). It has been reported that following the harvest of legume crops, a significant quantity of residual nitrogen remains available for subsequent crops (Banik et al., 2018), contributing to increased soil organic matter and serving as a cost-effective nutrient source for succeeding cropping seasons (Bhowmik & Das, 2018). Non-legume crops grown after legumes frequently exhibit enhanced yield and nutrient uptake efficiency due to better soil fertility and increased availability of both macro- and micronutrients (Chalk et al., 2014).

Upon decomposition, these residues release nitrogen gradually, improving soil fertility and enhancing nutrient cycling efficiency (Jensen et al., 2012). Moreover, legume-derived nitrogen has been shown to reduce dependence on synthetic fertilizers, thereby lowering production costs and minimizing environmental impacts such as greenhouse gas emissions and nitrate leaching (Meena et al., 2018). In addition to nitrogen benefits, legume residues contribute to improved soil structure, microbial activity, and water-holding capacity, all of which support the productivity of subsequent crops (Kebede, 2021).

Legume Crops Preceding Cereal Crops in Egypt

In a comparative study implemented by Ouda et al. (2020) to evaluate the effect of three crop sequences contain legume crops preceding wheat on soil nitrogen content after the harvest of each crop in the crop sequences. These crop sequences were maize, fahl clover (a variety of Egyptian clover) then wheat (CS1); cowpea, fahl clover then wheat (CS2); and cowpea intercropped with maize, fahl clover then wheat (CS3). Fahl clover was used in that study as an improved fallow to increase soil nitrogen content. Table 6.1 presents the percentage increase in soil nitrogen content after the harvest of each crop in the three crop sequences in both growing seasons. The results showed that after the harvest of maize in CS1 the increase in soil nitrogen content was 15% in the first growing season as a result of the residual

Table 6.1 The percentage increase in soil nitrogen content after the harvest of each crop in the three crop sequences in both growing seasons

	First growing season				
	Maize	**Fahl clover**	**Wheat**	**Cowpea**	**Cowpea/maize**
Maize, fahl clover, wheat	15	63	25	–	–
Cowpea, fahl clover, wheat	–	125	88	100	–
Cowpea/maize, fahl clover, wheat	–	88	50	–	75
	Second growing season				
Maize, fahl clover, wheat	12	115	75	–	–
Cowpea, fahl clover, wheat	–	150	113	138	–
Cowpea/maize, fahl clover, wheat	–	125	88	–	113

Prepared by the authors from Ouda et al. (2020)

effect of the preceding faba bean planting before the execution of the experiment. Whereas in the second growing season, the percentage of increase in soil nitrogen after the harvest of maize was increased to 12%, as a result of wheat preceded by fahl clover planted in the first growing season. Fahl clover cultivated in CS1 increased soil nitrogen content by 63% and 115% in the first and second seasons, respectively, where it was used by the following wheat and the leftover after wheat harvest was increased 25% and 75%, in the first and second seasons, respectively.

Table 6.1 also showed that cowpea and fahl clover in CS2 raised soil nitrogen content by 125% and 150%, respectively, enhanced the growth and yield of the following wheat and increased soil nitrogen content after wheat harvest by 88% and 113% in the first and second season, respectively. In CS3, the increase in soil nitrogen content after the harvest of cowpea intercropping system was 100% and 113% in the first and second season, respectively. The residual soil nitrogen content after fahl clover and wheat in CS3 was increased by 88% and 50%, respectively, in the first growing season and it was 113% and 88% in the second growing season, respectively.

Comparison Between Legume Crops in Their Superiority in Fixing Atmospheric Nitrogen under Intercropping

It has been reported that faba bean has the highest reliance on N_2 fixation for growth in comparison with peas, which leads to high nitrogen credit for the following crops (Hauggaard-Nielsen et al., 2009). Faba bean has been shown to attain higher amounts of nitrogen derived from the atmosphere, than peas. In addition, faba bean can reduce NO_3-N leaching by scavenging residual soil NO_3-N, higher than pea (Peoples et al., 2009). Carlsson et al. (2017) reported that, in peas intercropping system with wheat, wheat is more competitive in its use of mineral nitrogen and urges peas to fix more atmospheric nitrogen by symbiosis to meet its needs, which hinders the development of weeds due to the lack of nitrogen resources. According to the findings of Rodriguez et al. (2020), the residual organic matter of peas intercropping system with wheat, which is proportionally richer in nitrogen compared to monocropping, can help to replenish the soil's mineral reserves and, thus, preserve its natural fertility. Moreover, peas exhibit a defined growth pattern, flowering and maturing sooner than wheat cultivars, with nitrogen being released from nodules post-flowering, coinciding with the nitrogen need during the grain filling stage of wheat (Carlsson et al., 2017). According to Carranca et al. (1999), intercropping wheat with faba beans resulted in the highest soil available N values because faba beans fix more nitrogen than chickpea and also because of the dense root systems of faba bean compared to chickpea (Shanmugam et al., 2022).

Comparative Studies Between Legumes in Their Superiority in Nitrogen Fixation in Egypt

A comparison between different legume crops in their ability to fix atmospheric nitrogen was conducted by Ouda and Zohry (2023). A two-year field experiment was conducted, in which three legume crops (faba bean, chickpea, and lentil) were intercropped with sugar beets. Three irrigation quantities were used: 100% crop evapotranspiration (ETc) as a control treatment, and 85% and 70% ETc as imposed water stress treatments were implemented, respectively. Their findings revealed an increase in soil-accessible nitrogen values under sugar beet intercropping systems with chickpea and faba by 13% and 14%, respectively, when compared to the value under sole sugar beet irrigated with the required amount. Their findings also revealed a larger value of soil-accessible nitrogen in the second season, which was caused by the residual influence of these legume crops.

Another study conducted in Egypt over the course of two growing seasons by Zohry and Ouda (2025) examined wheat intercropping systems with three legume crops: wheat, peas, and fahl clover. The crops were irrigated using three different treatments: required quantity (100% ETc) and two water stress amounts (85% ETc and 70% ETc). Soil accessible nitrogen values were found to be higher under the three wheat intercropping systems compared to sole wheat cultivation when the required quantity of irrigation was applied, according to their research. Decomposition of legume crop residue benefits the next wheat in the second growing season, which in turn increases the values of soil accessible N compared to the first season. The findings of this research indicated that the highest soil available nitrogen values were found when faba beans were intercropped with wheat (23%), compared to the values obtained when either peas (16%) or fahl clover (5%) was intercropped with wheat.

Furthermore, Koriem and Hamd-Alla (2025) evaluated the comparative efficiency of soybean and cowpea when intercropped with maize in terms of their capacity for atmospheric nitrogen fixation and subsequent contribution to soil fertility. Their findings revealed that both legumes significantly enhanced soil nitrogen availability following harvest; however, notable differences were observed between the two species. Specifically, soil available nitrogen increased by 104% after soybean harvest, whereas cowpea demonstrated an even greater contribution, resulting in a 146% increase. This superior performance of cowpea may be attributed to its higher biological nitrogen fixation efficiency, greater root biomass production, and more active rhizosphere interactions, which collectively enhance nitrogen inputs into the soil.

These results highlight the importance of legume species selection in intercropping systems, as variations in nitrogen fixation potential can substantially influence the magnitude of residual soil fertility benefits for subsequent crops.

The Impact of Legumes Planting Density under Intercropping on Fixed Atmospheric Nitrogen

Increasing legume planting density in intercrops generally enhances the total amount of biologically fixed nitrogen by intensifying competition with the companion crop for mineral soil nitrogen, thereby stimulating legumes to rely more on atmospheric N_2 fixation (Bedoussac et al., 2015). Similarly, a higher density of the companion (non-legume) crop can further promote fixation by depleting available soil nitrogen, which shifts the legume's nutrition toward symbiotic pathways. Intercropping systems, particularly under low-input conditions, often increase both legume N_2 fixation and the companion crop's uptake of soil nitrogen due to interspecific complementarity and facilitation (Rodriguez et al., 2020).

Beyond this general pattern, several studies have shown that optimizing legume density is critical for maximizing N_2 fixation efficiency and total nitrogen input at the system level. Higher legume densities typically lead to greater nodule formation, increased root biomass, and enhanced rhizodeposition, all of which contribute to elevated nitrogen fixation (Jensen et al., 2012). However, this response is not linear; excessively high legume densities may intensify intraspecific competition, reduce individual plant performance, and potentially limit fixation efficiency per plant (Hauggaard-Nielsen et al., 2009). Therefore, an optimal density exists where total fixed nitrogen per unit area is maximized without compromising plant vigor.

In cereal–legume intercrops, adjusting the relative densities of both components has been shown to regulate nitrogen partitioning, with denser cereal stands increasing the proportion of nitrogen derived from the atmosphere (%Ndfa) in legumes (Naudin et al., 2010). Moreover, plant density influences canopy structure and light interception, indirectly affecting photosynthetic activity and the carbon supply required for symbiotic N_2 fixation, a highly energy-demanding process (Barbieri et al., 2023).

Legumes Planting Density Under Intercropping Effects on Nitrogen Fixation in Egypt

In Egypt, few studies have demonstrated that legume planting density is a key factor influencing nitrogen dynamics in intercropping systems. Ouda and Zohry (2026) intercropped three soybean planting densities (50%, 65%, and 75% of its recommended planting density) with maize (100% of its recommended planting density) in a two-year field experiment. They reported an increase in the available soil nitrogen with increasing soybean planting density, where the highest values were obtained when soybean planting density was 75%.

Intercropping legumes such as faba bean, pea, and clover with wheat at varying densities has been shown to increase soil available nitrogen compared to sole cropping systems, reflecting enhanced biological nitrogen fixation (Zohry & Ouda, 2025). However, the magnitude of this contribution depends on legume density, where moderate densities often achieve a balance between maximizing N_2 fixation and minimizing interspecific competition. Furthermore, Abdel-Wahab and El Manzlawy (2016) intercropped faba bean (50% of its plant density) with wheat under different plant densities, namely 100, 83.3, and 67%. They reported that the highest nitrogen use efficiency was found when 50% faba bean was intercropped with 67% wheat, which implied the complementary impact of faba bean in fixing atmospheric N_2.

The Impact of Legume Planting Date Under Intercropping on Fixed Atmospheric Nitrogen

The planting date of a legume companion crop significantly affects the amount of atmospheric nitrogen fixed, primarily by influencing the competitive dynamics between the legume and the companion crop and the environmental conditions during the legume's critical growth and nodulation periods. It plays a crucial role in determining biological nitrogen fixation by influencing crop synchrony, competition, and environmental conditions during nodulation. When legumes are sown at an optimal or slightly earlier date relative to the companion crop, they establish faster, access more light, and rely more on atmospheric N_2 due to reduced soil nitrogen availability, thereby enhancing fixation (King & Hofmockel, 2017). In contrast, delayed sowing often reduces biological nitrogen fixation because legumes face stronger competition for light and resources from the already established companion crop, leading to lower biomass and nodule activity (Megawer et al., 2010). International studies indicate that nitrogen fixation and transfer in intercrops are highly dependent on crop growth dynamics and seasonal conditions, with nitrogen fixation generally increasing as legumes develop and interact with companion crops over time (Salgado et al., 2021). Moreover, timely establishment is essential for effective rhizobia symbiosis and nitrogen input, as environmental factors such as soil moisture and temperature at sowing strongly regulate nodulation success and overall fixation in intercropping systems (Barbieri et al., 2023). Overall, optimizing planting date in legume-based intercrops enhances total fixed nitrogen by improving early establishment, prolonging the fixation period, and ensuring better complementarity with associated crops.

Legumes Planting Dates Effects on Nitrogen Fixation under Intercropping in Egypt

In Egypt, limited studies have evaluated the effect of legume planting date in intercropping systems. Sheha et al. (2015) showed that advancing pea sowing relative to wheat under their intercropping system enhanced nitrogen uptake and use efficiency, indicating increased biological nitrogen fixation. They intercropped peas with wheat in three planting dates for peas: 6 weeks, 4 weeks, and 2 weeks before wheat cultivation, and they found that the highest total count of rhizobia in rhizosphere of wheat roots was higher under the three peas planting dates, compared to the values obtained under the sole cultivation of wheat with the highest value being found when peas were intercropped 6 weeks before wheat cultivation.

This response is consistent with findings from other Egyptian intercropping studies, where management practices that enhance legume competitiveness—such as higher planting density—also increase soil nitrogen availability and N fixation contribution. For example, El-Mehy et al. (2023) relay intercropped maize with common dry bean in three planting dates of maize, where the co-growth duration between maize and beans was 60 days, 45 days, and 30 days long. They found that soil content of available nitrogen obtained from the co-growth duration was increased by 42%, 45%, and 42%, respectively, compared to the value obtained from maize monoculture.

Water Stress Suppresses the Ability of Legumes to Fix Atmospheric Nitrogen

Water stress greatly affects nitrogen fixation, resulting in diminished activity and perhaps irreparable damage to nodules (Basal & Szabó, 2020). Moreover, water stress diminishes the activity of nitrogenase, the enzyme that facilitates nitrogen fixation, and reduces oxygen uptake in nodules (Lumactud et al., 2022). Barzana et al. (2021) reported a reduction in nutrient availability for plants under low soil water conditions, which in turn limits nutrient uptake and assimilation. However, intercropping has a facilitative role in mitigating stress through root interactions and mobilization of limited or unavailable nutrients under harsh environmental conditions (Latati et al., 2014). Furthermore, Bargaza et al. (2015) indicated that intercropping faba bean with wheat under low water availability promotes faba bean nodulation and encourages root growth into deeper soil layers.

In addition, the timing of legume establishment within intercropping systems can further influence nitrogen fixation under water-limited conditions by altering competitive dynamics and resource partitioning (Corre-Hellou et al., 2006). Early sowing of the legume component relative to the companion crop has been shown to improve access to soil moisture and light, thereby sustaining nodule formation and nitrogenase activity during critical growth stages, whereas delayed sowing often exacerbates drought stress due to shading and competition for water (Bedoussac et al., 2015). Improved temporal complementarity in intercrops can enhance water-use efficiency and maintain higher biological nitrogen fixation compared with monocultures under drought conditions (Lithourgidis et al., 2011). Therefore, optimizing planting date in legume-based intercropping systems is essential not only for maximizing resource-use efficiency but also for sustaining nitrogen fixation under water stress.

Legume-Intercropping Mitigates the Effect of Water Stress on Fixed Nitrogen in Egypt

Limited studies were found in Egypt on the effect of legume intercropping systems on biological nitrogen fixation. Intercropping three legume crops, namely faba bean, peas or fahl clover with wheat contributed to increasing soil available nitrogen under water stress treatments, compared to the values obtained under sole wheat cropping, with the highest values obtained under faba bean intercropping system with wheat (Zohry & Ouda, 2025). Similar trend was observed when soybean was intercropped with maize under moderate water stress, where the accessible values of soil nitrogen were higher than that of sole maize intercropping, thereby highlighting the advantages of intercropping leguminous crops like soybean with maize (Ouda & Zohry, 2026).

Deep Nutrient Recycling

Deep nutrient recycling by legumes primarily refers to their ability to draw nutrients from deeper soil layers and surface-apply them through leaf litter and root residue decomposition (Yuvaraj et al., 2020). The deep roots of legumes create channels (macropores) that improve soil aeration and water infiltration, which enhances the movement and retention of nutrients in the soil profile (Kebede, 2021). The increased organic matter and root exudates also stimulate the overall diversity and activity of beneficial soil microorganisms, which play a critical role in nutrient cycling processes like mineralization (Liang, 2013). This process, combined with their renowned nitrogen-fixing capabilities, enhances overall soil fertility and nutrient availability for subsequent or companion crops (Srivastava et al., 2025). According to Barzana et al. (2021), intercropping peas with cereals enhances production stability by belowground nutrient transfer via direct root contact, diffusion of exudates, and mycorrhizal associations.

Phosphorus Mobilization

Legumes mobilize phosphorus in the soil throughout their growth, enhancing phosphorus uptake in subsequent or companion cereal crops (Mat Hassan et al., 2012). Shen et al. (2024) documented an elevation in accessible phosphorus within the inter-root soils of a soybean-maize intercropping system. Ren et al. (2021) indicated that soybeans may release protons and facilitate the mobilization of insoluble soil phosphorus due to rhizosphere acidification through the exudation of malate and citrate, hence enhancing soil phosphorus availability and its uptake by maize.

Furthermore, it has been reported that the roots of chickpea can increase phosphatase enzyme activities in the soil more than other legume crops (Maseko & Dakora, 2013) and mobilize less-labile forms of phosphorus by exudation of oxalic, malic, and citric acids into available phosphorus (Mat Hassan et al., 2012). This ability of chickpea roots in mobilizing soil phosphorus due to having a variety of root mechanisms as explained by Li et al. (2003), as well as its developed root architecture (Vandamme et al., 2013). According to Kusiab et al. (2021), in chickpea intercropping system with sugar beet, the mobilization of phosphorus by chickpea roots enhances root growth in sugar beet, where phosphorus requirement of sugar beet is high, as it is necessary for plant growth, energy storage, and high yields (Sims, 2010).

Cuartero et al. (2022) similarly observed that, in cowpea intercropped with a melon system, there are significantly higher amounts of accessible phosphorus than in monocropping. Furthermore, the availability of phosphorus in the rhizosphere was enhanced in a cowpea intercropping system, accompanied by notable pH alterations (Latati et al., 2014). Ferguson et al. (2013) demonstrated that legumes are more adept at extracting calcium and magnesium from the soil than cereals, substituting these elements with hydrogen, which leads to the removal of OH^- ions and an increase in H^+, consequently lowering soil pH. Furthermore, intercropping cowpea with maize enhanced nutrient recycling and facilitated biological nitrogen fixing (Dimande et al., 2024).

The mobilization of soil organic phosphorus by legume species occurs via the release of carboxylates, the enhancement of phosphatase activity, and alterations in root morphology and physiology, which facilitate phosphorus acquisition by cereals in intercropping systems with legumes (Wang et al., 2025). Xue et al. (2016) reported that carboxylate concentrations in the faba bean rhizosphere were 10 to 20 times higher than those in wheat within the intercropping system, indicating a substantial capacity for soil phosphorus mobilization and improved phosphorus uptake by wheat, aided by faba bean. Guo et al. (2021) assert that intercropping peanuts with maize enhances beneficial soil bacteria, optimizes phosphonate and phosphinate metabolism, and increases phosphorus availability for maize. Furthermore, Chongloi and Sharma (2018) observed an enhancement in the levels of phosphorus in oats when intercropped with peas, in contrast to sole oat cultivation.

In Egypt, an increase in soil available phosphorus by 2% was reported by Abd-Allah et al. (2021) when faba bean was intercropped with sugar beet in a two-year field experiment, compared to the value obtained under sole sugar beet cultivation. Whereas, Koriem and Hamd-Alla (2025) indicated that cowpea has the ability to mobilize more soil phosphorus compared to soybean, where the percentages of increase in soil available phosphorus were 70% and 48%, respectively compared to the value exist before cultivation.

Micronutrients Mobilization by Legumes

Legumes can alter the soil pH in the rhizosphere, establishing a microenvironment conducive to nutrient availability for intercrops (Goyal & Habtewold, 2023). The legume-based intercropping systems enhanced the enzymatic activities of soil microorganisms (Kumar et al., 2021). Incorporating biological carbon and nitrogen using a legume-based intercropping system significantly enhanced soil enzyme activity and nutrient cycling, augmenting soil fertility (Li et al., 2020). In intercropping, plant roots exude enzymes, including phosphatases, phytase, and carboxylates (Rojas-downing et al., 2017), which enhance the mobilization and utilization of nutrients such as iron, manganese, and zinc (Marschner, 2012) through the decomposition of organic residues (Xiao et al., 2013). The subterranean direct mutual benefits in legume-cereal intercropping systems, not only enhanced nitrogen transfer to cereals, but also improved bioavailability of iron and zinc to legume crops (Stagnari et al., 2017). Chongloi and Sharma (2018) observed an enhancement in the levels of potassium, zinc, and iron in oats when intercropped with peas, in contrast to sole oat cultivation.

The Role of Legumes in Micronutrients Mobilization in Egypt

Several investigations have been conducted in Egypt to study the role of legumes in micronutrients mobilization. It has been reported that cowpea has the ability to enhance soil potassium level more than soybean, compared to the level existing before cultivation, where the percentages of increase in potassium level were increased by 37% and 26% for cowpea and soybean, respectively (Koriem & Hamd-Alla, 2025). An increase in the potassium level after the harvest of faba bean intercropping system with sugar beet by 7% was reported, compared to the value obtained under monocultivation of sugar beet (Abd-Allah et al., 2021). Furthermore, inclusion of legume crops in each summer and winter season in a three-year crop rotation, where legume crops were 50% of the total cultivated crops, resulted in an increase in the value of soil available nitrogen, phosphorus, and potassium by 43%, 45%, and 35%, respectively, at the end of rotation (Abou-Keriasha et al., 2012).

In a field study by Awad et al. (2023), the amount of soil available nitrogen, phosphorus, and potassium were assessed after cultivation of four different crop sequences (Table 6.2). They reported an increase in the soil nutrients after the harvest of these crop sequences, compared to their values before cultivation. The highest increase in soil nitrogen amount was found when soybean was intercropped with maize followed by fahl clover then wheat, i.e., 36%. The highest percentage of increase in available phosphorus was found when maize was followed by fahl clover, then wheat. Whereas the highest increase in soil potassium was found when soybean was intercropped with maize then wheat.

Table 6.2 Percentage of increase in soil available nitrogen, phosphorus, and potassium after cultivation of four different crop sequences

	Nitrogen	Phosphorus	Potassium
Maize, fahl clover, wheat (CS1)	14	4	26
Cowpea/maize, fahl clover, wheat (CS2)	34	3	8
Soybean/maize, fahl, wheat (CS3)	36	2	27
Soybean/maize, wheat (CS4)	25	0	32

Prepared by the authors from Awad et al. (2023)

Organic Matter Addition

Legume residues (roots, shoots, and leaves) have a relatively low carbon-to-nitrogen (C/N) ratio compared to many other plant residues, which facilitates rapid decomposition by soil microorganisms, thus speeds up the cycling and release of nutrients back into the soil in plant-available forms (Yadav et al., 2017). Legumes notably enhance soil quality by enriching it with nitrogen through atmospheric nitrogen fixation and organic matter from the incorporation of organic residues (Hufnagel et al., 2020). The symbiotic association of legumes with nitrogen-fixing Rhizopus in their root nodules renders them exceptional crops for enhancing soil organic matter (Nazir et al., 2024). Reports indicate that following the harvest of legume crops, a significant quantity of residual nitrogen is available for subsequent crops, contributing to the enrichment of soil fertility through organic matter (Banik et al., 2018).

Furthermore, Bargaza et al. (2015) proposed that intercropping with faba bean enables its roots to penetrate deeper soil layers, thereby accessing a greater soil volume (Belachew et al., 2019), which may contribute to increased organic matter decomposition in the soil (Arcand et al., 2014). Legumes enhance carbon sequestration in the soil via their extensive root systems and nitrogen-fixing capabilities, which promote the accumulation of soil organic matter, functioning as a carbon sink (Hossain et al., 2023). Furthermore, legumes enhance soil organic matter, hence enhancing soil structure, soil fertility, water retention, and overall soil health (Stagnari et al., 2017). Liu et al. (2024) report that soil organic matter in a legume–maize intercropping system significantly increased by 4–14% compared to monoculture cultivation of maize.

In Egypt, in a three-year crop rotation implemented by Abou-Keriasha et al. (2012), containing 50% legume crops, the soil organic matter level rose by 38%, compared to non-legumes crop rotation. Furthermore, Abbady et al. (2016) reported that a 17% increase in soil organic matter when peas were intercropped with wheat compared to sole wheat cultivation. Cultivation of soybean resulted in an increase in the soil organic matter by 85%, whereas cultivation of cowpea resulted in a higher increase in soil organic matter by 171%, compared to the value existing before cultivation (Koriem & Hamd-Alla, 2025).

Additionally, Zohry et al. (2020) intercropped 50% peas with 100% wheat of the recommended density in the winter season and 50% cowpea intercropped with 100% sunflower of the recommended density in the summer season. Their results

Table 6.3 Soil organic content (%) before and after cultivation of legume intercropping systems and percentage of increase

Soil depth (cm)	Before cultivation	Organic matter after		Percentage of increase (%)
		First season	Second season	
Peas–wheat intercropping				
0–10	1.88	2.10	2.24	12
10–30	1.75	1.98	1.88	13
30–60	1.69	1.75	1.78	4
Average	1.77	1.94	1.97	9
Cowpea–sunflower intercropping				
0–10	1.88	2.09	2.22	11
10–30	1.75	1.39	1.83	21
30–60	1.69	1.72	1.74	2
Average	1.77	1.73	1.93	11

Prepared by the authors from Zohry et al. (2020)

indicated an increase in soil organic matter by 9% for peas intercropping system with wheat and by 11% for cowpea intercropping system with sunflower after the harvest of these intercropping systems (Table 6.3).

Conclusion

Legume-based cropping systems represent a fundamental component of sustainable agriculture due to their capacity to enhance soil nutrient circulation and overall agroecosystem functioning. Through biological nitrogen fixation, legumes contribute substantial amounts of plant-available nitrogen to the soil, benefiting both companion and subsequent crops and reducing reliance on synthetic fertilizers. In addition to nitrogen enrichment, legumes improve soil structure, increase organic matter content, stimulate beneficial microbial populations, and enhance the mobilization of essential nutrients such as phosphorus, potassium, and micronutrients.

The integration of legumes into cropping systems through intercropping and crop rotation significantly improves nutrient-use efficiency, soil fertility, and system productivity. Field evidence demonstrates that legume inclusion increases soil available nitrogen, phosphorus, potassium, and organic matter, thereby strengthening soil health and supporting sustainable crop production. The magnitude of these benefits is influenced by management practices such as legume species selection, planting density, planting date, and irrigation management. While water stress can reduce nitrogen fixation and nutrient availability, intercropping systems involving legumes often mitigate these effects through complementary resource use and improved root interactions.

Overall, legume-based cropping systems provide a practical pathway for enhancing nutrient cycling, improving soil fertility, and promoting resilient agricultural systems. Their adoption can support sustainable intensification by increasing productivity while conserving natural resources and maintaining soil health. Future research should focus on optimizing management practices, assessing long-term soil fertility impacts, and evaluating economic feasibility to facilitate wider adoption of legume-based cropping systems in diverse agroecological environments, particularly under the challenging conditions of arid and semi-arid regions.

References

Abbady, K. A., El-Maaz, E. I. M., Ahmed, H. M. R. M., & Zohry, A. A. (2016). Carbon sequestration as a function of intercropping management practices and different nitrogenous fertilizer types. *Journal of Soil Science and Agriculture Engineering, 7*(8), 565–586.

Abbas, Z. M., Khalil, F. A. F., & Eletr, W. M. T. (2013). Influence of water regimes and soil conditioners on yield, yield components and water utilization of Egyptian clover (variety Fahl). *Journal of Plant Production, 4*(11), 1675–1690.

Abd-Allah, A. M. M., Mohamed, M. S., & Noreldin, T. (2021). Water deficiency relief by application of potassium silicate to faba bean intercropped with sugar beet in sandy soil. *Zagazig Journal of Agricultural Research, 148*(1), 1–18.

Abdel-Wahab, T. I., & El Manzlawy, A. M. (2016). Yield and quality of intercropped wheat with faba bean under different wheat plant densities and slow—Release nitrogen fertilizer rates in sandy soil. *American Journal of Experimental Agriculture, 11*(6), 1–22.

Abou-Keriasha, M. A., Mohamed, W. M. A., Eiasa, N. M., & Kamel, A. S. (2012). Intensive crop rotations to improve agricultural production in Middle Egypt. *Egyptian Journal of Agricultural Research, 90*(4), 427–443.

Akchaya, K., Parasuraman, P., Pandian, K., Vijayakumar, S., Thirukumaran, K., Mustaffa, M. R. A. F., Rajpoot, S. K., & Choudhary, A. K. (2025). Boosting resource use efficiency, soil fertility, food security, ecosystem services, and climate resilience with legume intercropping: A review. *Frontiers in Sustainable Food Systems, 9*, 1527256. https://doi.org/10.3389/fsufs.2025.1527256

Amossé, C., Jeuffroy, M. H., Mary, B., & David, C. (2014). Contribution of relay intercropping with legume cover crops on nitrogen dynamics in organic grain systems. *Nutrient Cycling in Agroecosystems, 98*, 1–14. https://doi.org/10.1007/s10705-013-9591-8

Arcand, M. M., Lemke, R., Farrell, R. E., & Knight, J. D. (2014). Nitrogen supply from below-ground residues of lentil and wheat to a subsequent wheat crop. *Biology and Fertility of Soils, 50*(3), 507–515.

Awad, M. M., El-Mehy, A., & Hassan, H. H. (2023). Impact of crop sequence and nitrogen fertilization levels on wheat (Triticum aestivum L.) productivity in Egypt. *Egyptian Journal of Agricultural Research, 101*(1), 83–94.

Banik, D., Upadhyay, H., Devi, B. P., & Darvhankar, M. (2018). Effect of different irrigation system on growth and development of Wheat-Chickpea intercropping. *Bulletin of Environmental Pharmacology and Life Sciences, 8*(1), 32–34.

Barbieri, P., Starck, T., Voisin, A. S., & Nesme, T. (2023). Biological nitrogen fixation of legumes crops under organic farming as driven by cropping management: A review. *Agricultural Systems, 205*, 103579. https://doi.org/10.1016/j.agsy.2022.103579

Bargaza, A., Isaacb, M. E., Jensena, E. S., & Carlsson, G. (2015). Intercropping of faba bean with wheat under low water availability promotes faba bean nodulation and root growth in deeper soil layers. *Procedia Environmental Sciences, 29*, 111–112. https://doi.org/10.1016/j.proenv.2015.07.188

Barzana, G., Rios, J. J., Lopez-Zaplana, A., Nicolas-Espinosa, J., Yepes-Molina, L., Garcia-Ibañez, P., et al. (2021). Interrelations of nutrient and water transporters in plants under abiotic stress. *Physiologia Plantarum, 171*, 595–619. https://doi.org/10.1111/ppl.13206

Basal, O., & Szabó, A. (2020). The combined effect of drought stress and nitrogen fertilization on soybean. *Agronomy, 10*, 384. https://doi.org/10.3390/agronomy10030384

Bedoussac, L., Journet, E.-P., Hauggaard-Nielsen, H., Naudin, C., Corre-Hellou, G., Jensen, E. S., Prieur, L., & Justes, E. (2015). Ecological principles underlying the increase of productivity achieved by cereal-grain legume intercrops in organic farming. A review. *Agronomy for Sustainable Development, 35*, 911–935. https://doi.org/10.1007/s13593-014-0277-7

Belachew, K. Y., Nagel, K. A., Poorter, H., & Stoddard, F. L. (2019). Association of shoot and root responses to water deficit in young faba bean (Vicia faba L.) plants. *Frontiers in Plant Science, 10*, 1063. https://doi.org/10.3389/fpls.2019.01063

Bhowmik, N. S., & Das, A. (2018). Biofertilizers: A sustainable approach for pulse production. In R. Meena, A. Das, G. Yadav, & R. Lal (Eds.), *Legumes for soil health and sustainable management* (pp. 445–485). Springer. https://doi.org/10.1007/978-981-13-0253-4_14

Blanchart, E., Villenave, C., Viallatoux, A., Barthès, B., Girardin, C., Azontonde, A. a., & Feller, C. (2016). Long-term effect of a legume cover crop (Mucuna pruriens var. utilis) on the communities of soil macrofauna and nematofauna, under maize cultivation, in southern Benin. *European Journal of Soil Biology., 42*, 136–144.

Borase D., N., Nath, C. P., Hazra K., K., Senthilkumar, M., Singh S., S., Praharaj C., S., et al. (2020). Long-term impact of diversified crop rotations and nutrient management practices on soil microbial functions and soil enzyme activity. *Ecological Indicators, 114*, 106322. https://doi.org/10.1016/j.ecolind.2020.106322

Boukar, O., Bhattacharjee, R., Fatokun, C., Kumar, P. L., & Gueye, B. (2013). Cowpea. In M. Singh, H. D. Upadhyaya, & I. S. Bisht (Eds.), *Genetic and genomic resources of grain legume improvement* (pp. 137–155).

Carlsson, G., Bedoussac, L., Cupina, B., Djordjevic, V., Gaudio, N., Jensen, E.-S., Jeuffroy, M.-H., Journet, E.-P., Justes, E., & Mikic, A. (2017). Does a mixture of pea varieties with different leaf morphology improve crop performance? In *Proceedings of the international conference on advances in grain legume cultivation and use, Novi Sad, Serbia, 27–28 September 2017.*

Carranca, C., de Varennes, A., & Rolston, D. (1999). Biological nitrogen fixation by faba bean, pea and chickpea, under field conditions, estimated by the 15N isotope dilution technique. *European Journal of Agronomy, 10*(1), 49–56. https://doi.org/10.1016/S1161-0301(98)00049-5

Chalk, P. M., Peoples, M. B., McNeill, A. M., Boddey, R. M., Unkovich, M. J., et al. (2014). Methodologies for estimating nitrogen transfer between legumes and companion species in agro-ecosystems: A review of 15N-enriched techniques. *Soil Biology and Biochemistry, 73*, 10–21. https://doi.org/10.1016/j.soilbio.2014.02.005

Chongloi, K. L., & Sharma, K. K. (2018). Quality attributes of pea and oat in an intercropping system in rice fallows as influenced by integrated nutrient management. *Journal of Pharmacognosy and Phytochemistry, SP1*, 1663–1667.

Corre-Hellou, G., Fustec, J. & Crozat, Y. (2006). Interspecific Competition for Soil N and its Interaction with N2 Fixation, Leaf Expansion and Crop Growth in Pea–Barley Intercrops. *Plant Soil, 282,* 195–208. https://doi.org/10.1007/s11104-005-5777-4.

Cuartero, J., Pascual, J. A., Vivo, J.-M., Özbolat, O., Sánchez-Navarro, V., et al. (2022). A first-year melon/cowpea intercropping system improves soil nutrients and changes the soil microbial community. *Agriculture, Ecosystems & Environment, 328*, 107856. https://doi.org/10.1016/j.agee.2022.107856

Deakin, W. J., & Broughton, W. J. (2009). Symbiotic use of pathogenic strategies: Rhizobial protein secretion systems. *Applied Soil Ecology, 7*, 312–320.

Deemy, J. B., Besterman, A. F., Hall, B. M., Tyler, K. N., & Takagi, K. K. (2022). Nutrient cycling. In T. Dalu & R. J. Wasserman (Eds.), *Fundamentals of Tropical Freshwater Wetlands* (pp. 133–160). Elsevier. https://doi.org/10.1016/B978-0-12-822362-8.00017-7

Dimande, P., Arrobas, M., & Rodrigues, M. Â. (2024). Intercropped maize and cowpea increased the land equivalent ratio and enhanced crop access to more nitrogen and phosphorus compared to cultivation as sole crops. *Sustainability, 16*, 1440. https://doi.org/10.3390/su16041440

El-Mehy, A. A., Shehata, M. A., Mohamed, A. S., Saleh, S. A., & Suliman, A. A. (2023). Relay intercropping of maize with common dry beans to rationalize nitrogen fertilizer. *Frontiers in Sustainable Food Systems, 7*, 1052392. https://doi.org/10.3389/fsufs.2023.1052392

Federer, W. T. (2012). Statistical design and analysis for intercropping experiments. In *Two crops. Springer Series in Statistics* (Vol. 1). Springer-Verlag.

Ferguson, B. J., Lin, M. H., & Gresshoff, P. M. (2013). Regulation of legume nodulation by acidic growth conditions. *Plant Signaling & Behavior, 8*(3), e23426.

Gairola, A., Kumar, V., Sharma, D., Kumar, S., & Rathore, S. S. (2025). Ecosystem Services in Agro-Ecosystems: A step towards sustainability. In S. Babu, A. Das, S. S. Rathore, R. Singh, & R. Singh (Eds.), *Ecological solutions to agricultural land degradation*. Springer. https://doi.org/10.1007/978-981-96-3392-0_15

Goyal, R. K., & Habtewold, J. Z. (2023). Evaluation of legume-rhizobial symbiotic interactions beyond nitrogen fixation that help the host survival and diversification in hostile environments. *Microorganisms, 11*(6), 1454. https://doi.org/10.3390/microorganisms11061454

Guo, F., Wang, M., Si, T., Wang, Y., Zhao, H., Zhang, X., et al. (2021). Maize-peanut intercropping led to an optimization of soil from the perspective of soil microorganism. *Archives of Agronomy and Soil Science, 67*(14), 1986–1999.

Hauggaard-Nielsen, H., Mundus, S., & Jensen, E. S. (2009). Nitrogen dynamics following grain legumes and subsequent catch crops and the effects on succeeding cereal crops. *Nutrient Cycling in Agroecosystems, 84*, 281–291.

Homulle, Z., George, T. S., & Karley, A. J. (2022). Root traits with team benefits: Understanding belowground interactions in intercropping systems. *Plant and Soil, 471*, 1–26. https://doi.org/10.1007/s11104-021-05165-8

Hossain, M. Z., Anawar, H. M. D., & Chaudhary, D. R. (2023). Legume plants in the context of global climate change: Challenges and scopes for environmental sustainability. In *Climate change and legumes*. Taylor & Francis Books. https://doi.org/10.1201/9781003214885

Hufnagel, J., Reckling, M., & Ewert, F. (2020). Diverse approaches to crop diversification in agricultural research. A review. *Agronomy for Sustainable Development, 40*. https://doi.org/10.1007/s13593-020-00617-4

Jensen, E. S., Peoples, M. B., Boddey, R. M., et al. (2012). Legumes for mitigation of climate change and the provision of feedstock for biofuels and biorefineries. A review. *Agronomy for Sustainable Development, 32*, 329–364. https://doi.org/10.1007/s13593-011-0056-7

Jhariya, M. K., Meena, R. S., & Banerjee, A. (2021). Ecological intensification of natural resources towards sustainable productive system. In *Ecological intensification of natural resources for sustainable agriculture*. Springer. https://doi.org/10.1007/978-981-33-4203-3_1

Kebede, E. (2021). Contribution, utilization, and improvement of legumes-driven biological nitrogen fixation in agricultural systems. *Frontiers in Sustainable Food Systems, 5*, 1–18. https://doi.org/10.3389/fsufs.2021.767998

King, A. E., & Hofmockel, K. S. (2017). Diversified cropping systems support greater microbial cycling and retention of carbon and nitrogen. *Agriculture, Ecosystems and Environment, 240*, 66–76.

Koriem, M., & Hamd-Alla, W. (2025). Impact of preceding crops and various fertilizer applications on sugar beet productivity, quality, and economic return. *Journal of Sohag Agriscience (JSAS), 10*(2), 182–200. https://doi.org/10.21608/jsasj.2025.464135

Kumar, A., Blagodaskaya, E., Dippold, M. A., & Temperton, V. M. (2021). Positive intercropping effects on biomass production are species-specific and involve rhizosphere enzyme activities: Evidence from a field study. *Soil Ecology Letters, 4*(4), 444–453. https://doi.org/10.1007/s42832-021-0108-0

Kumari, V. V., Balloli, S. S., Kumar, M., Ramana, D. B. V., Prabhakar, M., Osman, M., et al. (2024). Diversified cropping systems for reducing soil erosion and nutrient loss and for increasing crop productivity and profitability in rainfed environments. *Agricultural Systems, 217*, 103919. https://doi.org/10.1016/j.agsy.2024.103919

Kusiab, N. Y. O., Stevens, W. B., Sintim, H. Y., Garciae, A. G., & Mesbah, A. O. (2021). Phosphorus fertilization and enhanced efficiency products effects on sugar beet. *Industrial Crops and Products, 171*, 1. https://doi.org/10.1016/j.indcrop.2021.113887

Latati, M., Blavet, D., Alkama, N., Laoufi, H., Drevon, J.-J., et al. (2014). The intercropping cowpea-maize improves soil phosphorus availability and maize yields in an alkaline soil. *Plant and Soil, 385*(1–2), 181–191.

Li, L., Tang, C., Rengel, Z., & Zhang, F. (2003). Chickpea facilitates phosphorus uptake by intercropped wheat from an organic phosphorus source. *Plant and Soil, 248*, 297–303.

Li, C. J., Hoffland, E., Kuyper, T. W., Yu, Y., Zhang, C., Li, H., Zhang, F., & van der Werf, W. (2020). Syndromes of production in intercropping impact yield gains. *Nature Plants, 6*, 653–660. https://doi.org/10.1038/s41477-020-0680-9

Liang, S. (2013). *Short-term effects of legume winter cover crop management on soil microbial activity and particulate organic matter. (Thesis)* (p. 89). North Carolina State University.

Lithourgidis, A. S., Dordas, CA., Damalas, C. A., Vlachostergios DN. (2011). Annual intercrops: An alternative pathway for sustainable agriculture. *Aust J Crop Sci. 5*(4), 396–410.

Liu, Q., Sun, X., Wu, W., Liu, Z., Fang, G., & Yang, P. (2022). Agroecosystem services: A review of concepts, indicators, assessment methods and future research perspectives. *Ecological Indicators, 142*, 109218. https://doi.org/10.1016/j.ecolind.2022.109218

Liu, R., Yang, L., Zhang, J., Zhou, G., Chang, D., Chai, Q., & Cao, W. (2024). Maize and legume intercropping enhanced crop growth and soil carbon and nutrient cycling through regulating soil enzyme activities. *European Journal of Agronomy, 159*. https://doi.org/10.1016/j.eja.2024.127237

Lumactud, R. A., Dollete, D., Liyanage, D. K., Szczyglowski, K., et al. (2022). The effect of drought stress on nodulation, plant growth, and nitrogen fixation in soybean during early plant growth. *Journal of Agronomy and Crop Science, 209*(3), 345–354. https://doi.org/10.1111/jac.12627

Marschner, P. (2012). In P. Marschner (Ed.), *Mineral nutrition of higher plants* (3rd ed.). Elsevier Ltd.

Maseko, S. T., & Dakora, F. D. (2013). Rhizosphere acid and alkaline phosphatase activity as a marker of P nutrition in nodulated Cyclopia and Aspalathus species in the Cape fynbos of South Africa. *South African Journal of Botany, 89*, 289–295.

Mat Hassan, H., Marschner, P., McNeill, A., & Tang, C. (2012). Grain legume pre-crops and their residues affect the growth, P uptake and size of P pools in the rhizosphere of the following wheat. *Biology and Fertility of Soils, 48*, 775–785. https://doi.org/10.1007/s00374-012-0671-8

Meena, B. P., Fagodiya, R. K., Prajapat, K., Dotaniya, M., Kaledhonkar, M., Sharma, P., et al. (2018). Legume green manuring: An option for soil sustainability. In R. Meena, A. Das, G. Yadav, & R. Lal (Eds.), *Legumes for soil health and sustainable management* (pp. 387–408). Springer. https://doi.org/10.1007/978-981-13-0253-4_12

Megawer, E. A., Sharaan, A. N., & El-Sherif, A. M. (2010). Effect of intercropping patterns on yield and its components of barley, lupine or chickpea grown in newly reclaimed soil. *Egyptian Journal of Applied Science, 25*(9), 437–452.

Mrunalini, K., Behera, B., Chandana, C., Patnaik, G. P., Modi, R. U., Saraswat, A., Rathi, N., & Kumar, N. (2022). Legumes to reduce ecological footprints for climate-smart cropping systems. In *Advances in legumes for sustainable intensification* (pp. 403–420). Academic Press. https://doi.org/10.1016/B978-0-323-85797-0.00032-X

Naab, J. B., Chimphango, S. M. B., & Dakora, F. D. (2009). N2 fixation in cowpea plants grown in farmers' fields in the Upper West Region of Ghana, measured using 15N natural abundance. *Symbiosis, 48*, 37–46.

Naudin, C., Corre-Hellou, G., Crozat, S. P. Y., & Jeuffroy, M. (2010). The effect of various dynamics of N availability on winter pea–wheat intercrops: Crop growth, N partitioning and symbiotic N2 fixation. *Field Crops Research, 119*(1), 2–11. https://doi.org/10.1016/j.fcr.2010.06.002

Nazir, M. J., Li, G., Nazir, M. M., Zulfiqar, F., Siddique, K. H. M., Iqbal, B., & Du, D. (2024). Harnessing soil carbon sequestration to address climate change challenges in agriculture. *Soil and Tillage Research, 237*, 105959. https://doi.org/10.1016/j.still.2023.105959

Ouda, S., & Zohry, A. (2023). Increasing sustainability by intercropping legume crops with sugar beet under imposed water stress. *Irrigation and Drainage*, 1–26. https://doi.org/10.1002/ird.2884

Ouda, S. a., & Zohry, A. (2026). Optimizing resource utilization through intercropping soybeans (Glycine max (L.) Merr.) and maize (Zea mays L.) in Middle Egypt. *Acta Agriculturae Slovenica, 122*(1), 1–23.

Ouda, S., Zohry, A., & Abdel-Wahab, S. (2020). Sustainable intensive cropping to reduce irrigation-induced erosion: II. Changing cropping sequence under sprinkler irrigation practice. *Moroccan Journal of Agricultural Science., 1*(1), 26–33.

Peoples, M. B., Brockwell, J., Herridge, D. F., Rochester, I. J., Alves, B. J. R., Urquiaga, S., Boddey, R. M., Dakora, F. D., Bhattarai, S., Maskey, S. L., et al. (2009). The contributions of nitrogen-fixing crop legumes to the productivity of agricultural systems. *Symbiosis, 48*, 1–17.

Preissel, S., Reckling, M., Schläfke, N., & Zander, P. (2015). Magnitude and farm-economic value of grain legume pre-crop benefits in Europe: A review. *Field Crops Research, 175*, 64–79. https://doi.org/10.1016/j.fcr.2015.01.012

Quinones, M., Lucas, M., & Pueyo, J. (2022). Adaptive mechanisms make lupin a choice crop for acidic soils affected by aluminum toxicity. *Frontiers in Plant Science, 12*. https://doi.org/10.3389/fpls.2021.810692

Rana, K., Goyal, N., & Sharma, G. P. (2018). Staging stewards of agro-ecosystems in the ecosystem services framework. *Ecosystem Services, 33*, 89–101.

Ren, Y., Zhang, L., Yan, M. et al. (2021). Effect of sowing proportion on above- and below-ground competition in maize–soybean intercrops. *Sci Rep, 11*, 15760. https://doi.org/10.1038/s41598-021-95242-w.

Rodriguez, C., Carlsson, G., Englund, J.-E., Flöhr, A., Pelzer, E., Jeuffroy, M.-H., Makowski, D., & Jensen, E. S. (2020). Grain legume-cereal intercropping enhances the use of soil-derived and biologically fixed nitrogen in temperate agroecosystems. A meta-analysis. *European Journal of Agronomy, 118*, 126077.

Rojas-downing, M. M., Nejadhashemi, A. P., Harrigan, T., & Woznicki, S. A. (2017). Climate risk management climate change and livestock: Impacts, adaptation, and mitigation. *Climate Risk Management, xxx*, 1–19. https://doi.org/10.1016/j.crm.2017.02.001

Salgado, G. C., Ambrosano, E. J., Rossi, F., Otsuk, I. P., Ambrosano, G. M. B., Santana, C. A., Muraoka, T., & Trivelin, P. C. O. (2021). Biological N fixation and N transfer in an intercropping system between legumes and organic cherry tomatoes in succession to green corn. *Agriculture, 11*, 690. https://doi.org/10.3390/agriculture11080690

Shanmugam, S., Hefner, M., Pelck, J. S., Labouriau, R., & Kristensen, H. L. (2022). Complementary resource use in intercropped faba bean and cabbage by increased root growth and nitrogen use in organic production. *Soil Use and Management, 38*, 729–740. https://doi.org/10.1111/sum.12765

Sheha, A. M., Abdel-Wahab, T. I., & Abdel-Wahab, S. I. (2015). Maximizing nitrogen and land use efficiencies of intercropped wheat with pea under different pea sowing dates. *Journal of Plant Sciences, 3*(6), 358–371. https://doi.org/10.11648/j.jps.20150306.20

Shen, X., Liu, Y., Li, X., & Li, L. (2024). Effect of intercropping with legumes at different rates on the yield and soil physicochemical properties of Cyperus esculentus L. in arid land. *Frontiers in Plant Science, 15*, 1351843. https://doi.org/10.3389/fpls.2024.1351843

Sims, A. L. (2010). Sugar beet response to broadcast and starter phosphorus applications in the Red 483 River Valley of Minnesota. *Agronomy Journal, 102*, 1369–1378. 484. https://doi.org/10.2134/agronj2010.0099

Song, M., Li, J., Gao, L., & Tian, Y. (2023). Comprehensive evaluation of effects of various carbon-rich amendments on overall soil quality and crop productivity in degraded soils. *Geoderma, 436*, 116529. https://doi.org/10.1016/j.geoderma.2023.116529

Srivastava, R. K., Yetgin, A., & Srivastava, S. (2025). The role of legume roots in carbon sequestration, soil health enhancement, and salinity mitigation under climate change: A comprehensive review. *Soil and Tillage Research, 253*, 106656. https://doi.org/10.1016/j.still.2025.106656

Stagnari, F., Maggio, A., Galieni, A., et al. (2017). Multiple benefits of legumes for agriculture sustainability: An overview. *Chemical and Biological Technologies in Agriculture, 4*, 2. https://doi.org/10.1186/s40538-016-0085-1

Vandamme, E., Renkens, M., Pypers, P., Smolders, E., Vanlauwe, B., & Merckx, R. (2013). Root hairs explain P uptake efficiency of soybean genotypes grown in a P-deficient Ferralsol. *Plant and Soil, 369*, 269–282. https://doi.org/10.1007/s11104-012-1571-2

Walia, S. S., Rani, N., Ravisankar, N., Bhagat, R., & Kaur, T. a. K. K. (2025). Legume-based crop rotation sustain the soil biodiversity, fertility levels, productivity, and profitability: Evidence from a long-term study under Indian subtropical conditions. *Frontiers in Agronomy, 7*, 1681733. https://doi.org/10.3389/fagro.2025.1681733

Wang, L., Chen, X., Cheng, B., Yang, H., Gao, Y., Li, X., et al. (2025). Stage-specific phosphorus mobilization enhances phosphorus uptake in relay-intercropped soybean. *Agriculture, Ecosystems & Environment, 388*, 109660. https://doi.org/10.1016/j.agee.2025.109660

Xiao, X., Cheng, Z., Meng, H., Liu, L., Li, H., & Dong, Y. (2013). Intercropping of green garlic (Allium sativum L.) induces nutrient concentration changes in the soil and plants in continuously cropped cucumber (Cucumis sativus L.) in a plastic tunnel. *PLoS One, 8*(4). https://doi.org/10.1371/journal.pone.0062173

Xu, Z., Li, C., Zhang, C., Yu, Y., van der Werf, W., & Zhang, F. (2020). Intercropping maize and soybean increases efficiency of land and fertilizer nitrogen use; A meta-analysis. *Field Crops Research, 246*, 107661.

Xue, Y., Xia, H., Christie, P., Zhang, Z., Li, L., & Tang, C. (2016). Crop acquisition of phosphorus, iron, and zinc from the soil in cereal/legume intercropping systems: A critical review. *Annals of Botany, 117*, 363–377. https://doi.org/10.1093/aob/mcv182

Yadav, G. S., Lal, R., Meena, R. S., Babu, S., Das, A., Bhomik, S. N., Datta, M., Layak, J., & Saha, P. (2017). Conservation tillage and nutrient management effects on productivity and soil carbon sequestration under double cropping of Rice in north eastern region of India. *Ecology in India*. http://www.sciencedirect.com/science/article/pii/S1470160X17305617

Yuvaraj, M., Pandiyan, M., & Gayathri, B. (2020). Role of legumes in improving soil fertility status. In *Legume crops—Prospects, production and uses*. Intechopen. https://doi.org/10.5772/intechopen.93247

Zabala, J. A., Martínez-Paz, J. M., & Alcon, F. (2021). Comprehensive approach for agroecosystem services and disservices valuation. *Science of the Total Environment, 768*, 1–14.

Zhang, M. M., Wang, N., Hu, Y. B., & Sun, G. Y. (2018). Changes in soil physicochemical properties and soil bacterial community in mulberry (Morus alba L.)/alfalfa (Medicago sativa L.) intercropping system. *Microbiology Open, 7*(2), e00555. https://doi.org/10.1002/mbo3.555

Zhao, Y., Tian, Y., Li, X., Song, M., Fang, X., Jiang, Y., & Xu, X. (2022). Nitrogen fixation and transfer between legumes and cereals under various cropping regimes. *Rhizosphere, 22*, 100546. https://doi.org/10.1016/j.rhisph.2022.100546

Zohry, A., & Ouda, S. (2025). Legume crops enhance water use efficiency under intercropping system with wheat. *Moroccan Journal of Agricultural Sciences, 6*(3), 145–160.

Zohry, A., Ouda, S., & Abdel-Wahab, T. (2020). Sustainable intensive cropping to reduce irrigation-induced erosion: Intercropping systems under surface irrigation practice. *Moroccan Journal of Agricultural Science, 1*(2), 63–71.

Chapter 7
Legume-Based Cropping System Achieves Biological Control

Abstract This chapter highlights the pivotal role of legume-based cropping systems in enhancing biological control within agroecosystems. It demonstrates how integrating legumes through intercropping, crop rotation, and cover cropping increases biodiversity, improves habitat complexity, and promotes the abundance and effectiveness of natural enemies such as predators and parasitoids. The chapter synthesizes ecological mechanisms underlying pest suppression, including host disruption, resource dilution, and enhanced natural enemy performance. Special emphasis is placed on key systems such as faba bean-based and cowpea–maize intercropping, which have shown significant reductions in insect pest incidence and improvements in beneficial insect populations. Additionally, the chapter explores the role of legumes in weed suppression through competitive interactions, and canopy shading, highlighting their contribution to integrated weed management. Evidence from Egyptian agroecosystems is presented to demonstrate the practical relevance of these systems under arid and semi-arid conditions, where legumes contribute to soil fertility, water-use efficiency, and ecological resilience. Overall, legume-based systems are positioned as a cornerstone of sustainable intensification, reducing reliance on chemical inputs while enhancing productivity and ecosystem services.

Keywords Faba bean · Cowpea · Insect pests · Beneficial insects · Weed suppression · Intercropping · Legume cover crops

Introduction

Biological control is the management of pests and weeds using natural enemies—predators, parasites, or pathogens—to reduce their populations to manageable levels. It acts as an environmentally sustainable, often self-sustaining alternative to chemical pesticides (Stenberg et al., 2021). By diversifying agricultural landscapes through intercropping, rotations, and ground cover systems, legumes disrupt pest and pathogen life cycles, improve soil health, and attract beneficial insects that

A. E. Zohry, S. A. H. Ouda, *The Agroecosystem Services of Leguminous Crops*,
https://doi.org/10.1007/978-3-032-32461-0_7

contribute to natural pest control (Lu et al., 2014). In conservation biological control strategies, increased plant diversity is consistently associated with greater abundance and diversity of beneficial insects, including natural enemies (Foyer et al., 2016). Rather than relying on chemical inputs, they function as an ecological foundation that supports a balanced, biodiverse, and resilient agroecosystem, thereby contributing to the biological control of pests, diseases, and weeds (Letourneau et al., 2011).

Legumes play a pivotal role in biological control within agroecosystems by enhancing the abundance, diversity, and activity of natural enemies of crop pests, thereby supporting ecologically based pest management (Pierre et al., 2021). Their integration into cropping systems—through practices such as intercropping, crop rotation, and cover cropping—enhances both the structural and functional diversity of habitats. This, in turn, promotes populations of predators and parasitoids that help regulate major agricultural pests (Pierre et al., 2023). In addition, legumes such as faba bean produce abundant flowers that serve as valuable sources of nectar and pollen for pollinators and natural enemies (Köpke & Nemecek, 2010). Legume-based cropping systems hold significant potential for advancing ecological intensification and developing low-emission, climate-resilient agricultural systems (Snapp et al., 2019). These systems can slow pest population buildup and reduce crop losses caused by pests (Cole et al., 2022).

Intercropping—the simultaneous cultivation of two or more crop species in the same field—is one of the most effective approaches for increasing biodiversity in agroecosystems (Li et al., 2013; Gu et al., 2021). Numerous studies have demonstrated that intercropping enhances the diversity and abundance of beneficial entomofauna (Pierre et al., 2022). Several ecological hypotheses explain this effect. First, increased plant diversity attracts more predators and parasitoids, thereby enhancing natural pest suppression (Vamdermeer, 1989). Second, monocropping facilitates host location for herbivorous insects, especially those with narrow host ranges, enabling them to feed and reproduce more efficiently (Root, 1973). Third, the presence of non-host plants may interfere with herbivores' ability to locate their preferred hosts (Finch & Collier, 2000). Finally, intercropping can alter host plant quality and reduce its suitability for herbivores compared to monocultures (Bach, 1981).

Intercropping also represents an effective strategy for weed management. By providing greater ground cover, crops suppress weed emergence and growth, which otherwise significantly reduce crop productivity and yield. As such, intercropping can serve as a viable alternative to herbicide use (Singh et al., 2017). It has been shown to limit the development of late-emerging weeds through shading effects (Takim, 2012). Weeds can consume up to 30–50% of applied fertilizers and reduce soil water availability by 20–40%, leading to substantial yield losses that may reach 100% under severe infestation (Abdulkareem et al., 2024; Gu et al., 2021).

Legume-based intercropping systems are particularly effective in suppressing weeds, reducing weed growth by 61–87% compared to sole cropping systems, primarily due to enhanced shading and competition for resources (Sharma & Banik, 2013). For example, legume–maize intercropping systems have been shown to reduce weed density and biomass compared to monoculture maize (Zohry, 2005). These systems provide both spatial and temporal competition, making them more effective than sole cropping in controlling weeds (Abou-Keriasha et al., 2012). The rapid canopy development of legumes in systems such as cowpea–maize, green gram–maize, and pea–cereal combinations effectively suppress weed emergence (Gitari et al., 2020). Similarly, Rahimi et al. (2019) reported that maize–black gram intercropping (1:1 or 2:2 ratios) significantly reduced weed density compared to sole maize. Shah et al. (2011) further observed that maize–soybean intercropping exhibited higher weed suppression efficiency than maize–green gram systems, primarily due to reduced availability of space and light, which limits weed growth.

Crop rotations that incorporate legumes also play a crucial role in pest management. These systems disrupt pest life cycles and reduce population buildup by introducing temporal diversity in host availability (Al-Musawi et al., 2025; Guinet et al., 2023; Zou et al., 2024). Alternating between legumes and non-legumes breaks host continuity for specialized pests, thereby reducing infestation levels and limiting reproduction (Bowles et al., 2020; Guinet et al., 2023). Moreover, diversified rotations enhance agroecosystem biodiversity, increasing the abundance of pollinators and natural enemies such as predators and parasitoids, which contribute to biological pest control (Ouyang et al., 2020; Pierre et al., 2023). Legume-based rotations also strengthen ecological regulation processes that suppress pests, diseases, and weeds, ultimately reducing reliance on synthetic pesticides and improving the sustainability of agricultural systems (Guinet et al., 2023; Ediagbonya et al., 2025).

Legumes are valuable cover crops that provide multiple agronomic benefits, including biological nitrogen fixation, soil improvement, and effective weed suppression. During fallow periods, in particular, they limit weed emergence by competing for resources and creating shading that inhibits weed growth (Kocira et al., 2020). Beyond weed control, legume cover crops can also influence pest dynamics. In some cases, they act as alternative hosts for insect pests, which can disrupt pest life cycles and ultimately reduce infestation levels (Dingha et al., 2021). Moreover, legumes contribute to pest management indirectly by enhancing soil health and crop vigor, thereby improving plant tolerance to pest damage. Improvements in soil structure, increased microbial activity, and greater nitrogen availability associated with legume systems promote stronger, more resilient crops that are less vulnerable to pest outbreaks (Rodríguez et al., 2022). Altogether, these mechanisms underscore the multifunctional role of legumes in supporting biological control, reducing reliance on synthetic pesticides, and advancing sustainable pest management strategies.

This chapter aims to examine the role of legume-based cropping systems in enhancing biological control of insect pests and to elucidate the ecological mechanisms through which legumes promote natural enemy and suppress pest populations. It evaluates the effectiveness of legume intercropping systems, particularly those involving faba bean and cowpea, as key leguminous crops, in reducing pest incidence and improving agroecosystem functioning. The chapter also assesses the contribution of legume-based systems to weed suppression and integrated weed management through competitive and ecological interactions. Furthermore, it highlights the relevance and applicability of these systems in Egyptian agroecosystems, especially under arid and resource-limited conditions. Ultimately, the chapter seeks to demonstrate how the integration of legumes can support sustainable intensification and climate-resilient agricultural production systems.

Legumes as Functional Components of Agroecosystems

Legume species serve as vital ecological infrastructure in agricultural systems by providing floral resources such as nectar and pollen, alternative prey, and structural shelter that sustain beneficial arthropods—including pollinators and natural enemies—throughout the growing season (Wyckhuys et al., 2023). By supplying consistent food resources, legumes enhance biodiversity and improve the longevity and reproductive capacity of predators (Rosa-Schleich et al., 2019). The abundant flowering of legumes produces sugar-rich nectar and protein-rich pollen, which support pollinators such as bees and hoverflies, as well as predatory arthropods (Cole et al., 2022).

In addition, legumes host non-pest insects that serve as alternative prey, helping to maintain populations of natural enemies even when pest densities are low. Their dense foliage provides shelter and, in some cases, suitable sites for oviposition (egg-laying), further enhancing habitat suitability for beneficial insects (Kirsch et al., 2023). By offering a continuous and diverse supply of resources, legumes sustain high populations of predators, parasitoids, and pollinators, thereby strengthening natural pest control and improving pollination services within cropping systems (Wyckhuys et al., 2023).

The Role of Faba Bean in Increasing Beneficial Insects

Open-nectary legumes such as faba bean are particularly effective in enhancing the survival and activity of beneficial insects. Faba bean is recognized as an important ecological component of agricultural systems due to its prolific flowering and high-quality floral resources (Ruhanen et al., 2024). It plays a key role in sustaining both pollinators and natural enemies by providing a reliable supply of protein-rich pollen and energy-rich nectar, especially during periods when alternative floral resources are limited (Suso et al., 2016; Chau et al., 2019; Cole et al., 2022). In addition to supporting beneficial insects directly, faba bean also attracts certain pest species, such as aphids, *Liriomyza leafminers*, and caterpillars, which serve as prey for natural enemies including ladybirds, lacewings, *Orius* spp., and parasitoids (Mohamed et al., 2023). This interaction helps sustain natural enemy populations and enhances biological control within the system.

Furthermore, intercropping faba bean with crops such as cabbage has been shown to improve the biological control of pests like *Plutella xylostella*, as the extrafloral nectaries (EFNs) of faba bean provide additional resources for natural enemies (Suso et al., 2016). Beyond its ecological benefits, faba bean also functions as a break crop, contributing to increased biodiversity while providing a sustainable, protein-rich source of food and feed (Ruhanen et al., 2024).

Cowpea–Maize Intercropping System Reduces the Incidence of Insect Pests

Cowpea–maize intercropping is an effective strategy for reducing insect pest incidence in maize by increasing habitat diversity, enhancing the activity of natural enemies, and disrupting pest colonization. In a field study, Rao et al. (2025) evaluated a 2:1 maize–cowpea intercropping system and monitored insect pest incidence through weekly assessments of larval populations, leaf damage, and natural enemy abundance (including coccinellids and spiders). Their results showed significantly lower larval infestation in intercropped systems compared to sole maize, with reductions of approximately 30% during the vegetative stage and 25% during the reproductive stage.

Fall armyworm (*Spodoptera frugiperda*), one of the most destructive pests of maize (Zacarias, 2020), can be effectively suppressed through intercropping with cowpea (Soujanya et al., 2024). Similarly, Pierre et al. (2022) demonstrated that maize–cowpea intercropping enhances both crop yield and the abundance of beneficial insects without increasing interspecific competition. Consistent findings were reported by Mukanga et al. (2024), who observed a significant reduction in fall armyworm infestation—ranging from 15.9% to 34.6%—in maize intercropped with cowpea compared to monoculture systems, which resulted in lower infestation rates, reduced leaf and cob damage, fewer larvae, and decreased ear rot incidence. In addition, a diverse community of natural enemies was recorded, including eight predatory insect species and eight larval parasitoids, with parasitism rates ranging from 0.85% to 6%.

Relevance to Egyptian Agroecosystem

In Egypt, several studies have demonstrated the effectiveness of cowpea–maize intercropping in reducing insect pest populations under local conditions. Abdel-Wahab et al. (2021), working under Middle Egypt conditions, reported that intercropping cowpea with maize—combined with recommended irrigation levels and appropriate spatial arrangements—significantly reduced infestations of aphids, jassids, and cotton leafworm in maize. The authors concluded that this intercropping system has strong potential to enhance biological pest regulation in Egyptian agroecosystems. In Upper Egypt, Hamd-Alla et al. (2014) intercropped cowpea with maize and reported that cowpea, when intercropped with maize, attracts beneficial insects, such as *Coccinella septempunctata*, *Coccinella undecimpunctata*, and

Chrysopa pallens. These insects prey on aphids that infest maize tassels, resulting in significant maize yield losses.

Similarly, Abd-Elsamed et al. (2011), in a study conducted in the Nile Delta, evaluated the impact of cowpea–maize intercropping on piercing–sucking insect populations. Their results showed that a 3:3 intercropping system significantly reduced the population densities of several key pests, including *Rhopalosiphum maidis*, *Rhopalosiphum padi*, *Aphis gossypii*, *Empoasca decedens*, *Empoasca decipiens*, *Cicadulina chinai*, and the whitefly *Bemisia tabaci*.

The Role of Legume-Based Intercropping System in Weed Suppression

Weeds not only reduce crop yields but also deteriorate grain quality (Plaza et al., 2022) and can serve as alternative hosts for pests and diseases (Kumar et al., 2021). Legume-based intercropping systems have been widely reported to suppress weed growth, particularly when combined with cereals such as maize or wheat (Sharma et al., 2021). This suppressive effect is mainly attributed to increased competition for essential resources, including light, water, and nutrients, especially nitrogen (Kemper et al., 2020).

Moreover, legumes often establish an early and dense canopy that shades the soil surface, limiting light availability for weed emergence. This shading reduces weed seed germination and growth, a mechanism commonly referred to as the *smothering effect* (Lithourgidis et al., 2011). Empirical evidence supports these findings, where Singh and Kumar (2023) demonstrated that intercropping maize with green gram or black gram significantly reduced weed density and weed biomass compared with maize monoculture. Similarly, Leoni et al. (2022) reported that relay intercropping of legumes with wheat is an effective ecological approach for weed control, particularly in low-input systems. Their study showed that weed suppression can reach up to 90% during the cropping season and 94% in the following spring.

The effectiveness of the *smothering effect*, however, depends strongly on the legume species and its biomass production. Perennial legumes, such as *Medicago sativa* and *Trifolium repens*, provide more consistent and long-term weed suppression by maintaining dense ground cover even after crop harvest (Vrignon-Brenas et al., 2016). In contrast, annual and some self-seeding legumes are less reliable and may, in some cases, promote weed growth due to insufficient soil coverage or nitrogen release. Importantly, the benefits of legume relay intercropping are less pronounced in high-input systems, where herbicides and strong crop competition already provide effective weed control. Therefore, this strategy is particularly valuable as a component of integrated weed management in low-input and organic farming systems (Leoni et al., 2022).

Recent studies highlight that the effectiveness of intercropping systems in suppressing weeds depends on several key agronomic factors. These include crop proportion, as higher proportions of the companion crop generally led to stronger weed suppression (Bedoussac et al., 2015; Lebreton et al., 2024). In addition, crop varieties with rapid early vigor and extensive canopy development are more effective at limiting weed growth (Jensen et al., 2020). Recent agroecological research suggests that optimizing these factors can substantially enhance the weed-suppressive potential of intercropping systems, making them a more efficient and sustainable strategy for weed management (Sammama et al., 2023).

Cowpea–Maize Intercropping System Suppresses Weed Growth

Reduced weed growth in cowpea–maize intercropping systems compared to sole maize cropping has been widely reported for enhancement of ground cover and intensification of competition for essential resources, thereby limiting weed establishment and growth (Jamshidi et al., 2013). For instance, Saudy (2015) showed that cowpea–maize intercropping planted on alternating ridges reduced weed biomass by approximately 50% compared to sole maize cultivation. Similarly, cowpea can act as a living mulch, decreasing weed biomass by up to 45.5% while effectively suppressing weed density and growth. This suppression is largely attributed to reduced light penetration and nutrient availability for weeds, which also contributes to improved land equivalent ratios (Eskandari & Kazemi, 2011; Kazemi et al., 2025). In addition, Verret et al. (2017) reported that intercropping cowpea with maize reduced weed biomass by 42% compared with a weeded control treatment and by 56% relative to a non-weeded control treatment, without negatively affecting maize yield. Overall, numerous studies indicate that cowpea–maize intercropping can substantially reduce weed biomass and minimize yield losses caused by weeds, making it an effective and sustainable weed management strategy (Katsaruware & Manyanhaire, 2009; Jamshidi et al., 2013; Suhi et al., 2022).

Relevance to Egyptian Agroecosystem

In Egypt, several studies have demonstrated the effectiveness of cowpea–maize intercropping in reducing weed growth. Zohry (2005) conducted a two-year field experiment and reported significantly lower weed biomass under intercropping compared to sole maize cultivation. Similarly, Abou-Keriasha et al. (2011) evaluated different planting dates for cowpea in association with maize—three weeks before maize, simultaneously with maize, and three weeks after maize. Their results showed that sowing cowpea at the same time as maize resulted in the lowest fresh weight of annual weeds. In a subsequent study, Abou-Keriasha et al. (2013) examined the combined effects of preceding winter crops and intercropping systems on weed dynamics in maize. The preceding crops included Egyptian clover, faba bean, and onion, alongside cowpea–maize intercropping. The authors found that the lowest weed biomass was consistently recorded in the cowpea–maize intercropping system after Egyptian clover and faba bean compared with maize grown as a monoculture preceded with the studied crops.

Effect of Faba Bean Intercropping Systems on Weeds Suppression

A growing body of evidence indicates that intercropping faba bean with cereals significantly enhances weed suppression compared with monocropping systems. In a recent multi-season field study, Villegas-Fernández et al. (2024) reported that faba bean–barley intercropping markedly reduced weed coverage and biomass by 92.7% and 76.6%, respectively. In contrast, faba bean–wheat intercrops decreased weed biomass by 46.1% without significantly affecting weed coverage, while faba bean–pea mixtures showed no significant weed suppression. These results demonstrate that the effectiveness of intercropping in controlling weeds varies considerably depending on species combinations and management practices. The limited performance of faba bean–pea intercrops, as observed by Villegas-Fernández et al. (2024), is likely due to the similar growth habits and resource requirements of the two legume species. This finding highlights the importance of functional diversity in the design of intercropping systems.

These observations are consistent with both earlier and recent studies that emphasize the superior weed–suppressive capacity of cereal–legume intercrops. For example, Dhima et al. (2018) reported that faba bean–barley systems reduced weed biomass by up to 50–99% compared with sole cropping. More recent studies, including Lebreton et al. (2024) and Bedoussac et al. (2015), further confirm that cereal–legume intercrops consistently outperform monocultures in reducing weed density and biomass across diverse agroecological conditions. In addition, Jensen et al. (2020) highlighted that combining cereals with legumes enhances crop competitiveness and improves weed control through complementary resource use.

Faba Bean Intercropping for Weed Management in Egypt

Faba bean intercropping systems have been widely investigated in Egypt as a key ecological tool for managing broomrape infestation in faba bean fields. Several studies demonstrate that introducing companion crops can significantly reduce parasite incidence, biomass, and seed bank replenishment. A study conducted under Egyptian field conditions showed that intercropping faba bean with fenugreek or garlic significantly reduced broomrape infestation while improving crop yield and plant survival (El-Mehy et al., 2022).

More recent work demonstrated that intercropping faba bean with flax, fenugreek, or radish significantly reduced broomrape biomass, length, and seed production. In particular, flax-based intercropping systems achieved the greatest reduction in infestation levels and soil seed bank, with reductions exceeding 50% in some cases (Hegazi et al., 2024). These findings confirm that intercropping is an effective strategy for suppressing parasitic weeds under Egyptian conditions.

The Agroecosystems Services of Crop Rotation in Biological Control

Legume-based crop rotations play a pivotal role in regulating pest populations and enhancing beneficial biodiversity within agroecosystems. By alternating legumes with non-legume crops, these rotations introduce temporal diversity that disrupts the life cycles of many insect pests, particularly those with narrow host ranges (Weisberger et al., 2019). The absence of a continuous host prevents pests from completing successive generations, thereby reducing their survival, reproduction, and overall population buildup (Bowles et al., 2020). This break in host continuity also limits the carryover of pests and associated pathogens between cropping seasons, making rotations an effective preventive strategy rather than a reactive control measure (Teasdale, 2017). In addition, legumes can alter soil conditions and crop nutritional profiles, which may further influence pest dynamics by reducing host plant suitability for herbivorous insects (Guinet et al., 2023).

Beyond pest suppression, legume-based rotations contribute significantly to increasing the abundance and effectiveness of pollinators and natural enemies (Zhang et al., 2019). Legumes provide essential floral resources such as nectar and pollen, which sustain beneficial arthropods including bees, hoverflies, parasitoids, and predatory insects (Pierre et al., 2023). These resources are particularly important during periods when other crops are not flowering, ensuring the continuous presence of beneficial organisms within the system. Furthermore, legume-based rotations can support alternative prey or hosts, allowing predator and parasitoid populations to persist even when pest densities are low. This ecological continuity enhances the stability and resilience of biological control processes (Gurr et al., 2017).

Legume-based rotations also improve habitat complexity both above and below ground. Aboveground, diversified cropping sequences create a more heterogeneous environment that supports a wider range of species and ecological interactions (Drinkwater et al., 1998). Belowground, legumes enhance soil health through biological nitrogen fixation, increased organic matter inputs, and stimulation of microbial activity (Weiner et al., 2024). These improvements contribute to stronger, more vigorous crops that are better able to tolerate or resist pest attacks. Moreover, healthy soils can support beneficial soil organisms that play a role in suppressing soil-borne pests and diseases (Berendsen et al., 2012; Pineda et al., 2017).

Collectively, these mechanisms demonstrate that legume-based crop rotations are a cornerstone of ecological intensification. They not only reduce pest pressure through life cycle disruption but also enhance ecosystem services by fostering diverse communities of pollinators and natural enemies. As a result, such systems reduce reliance on chemical inputs, improve agroecosystem resilience, and support the development of sustainable and climate-smart agricultural practices (Snapp et al., 2019; Letourneau et al., 2011).

The Role of Legumes as Cover Crops

Cover crops are characterized as non-cash crops cultivated prior to or concurrently with primary crops to maintain soil coverage with vegetation for extended periods (Blanco-Canqui et al., 2015). Leguminous cover crops play a multifaceted role in agroecosystems by indirectly reducing pest pressure through improvements in soil health, crop vigor, and ecological balance. One of the primary mechanisms underlying this effect is the enhancement of soil fertility, particularly through biological nitrogen fixation, which increases nitrogen availability for subsequent crops. Improved nutrient status promotes vigorous plant growth, enabling crops to better tolerate and recover from insect herbivory and pathogen attack (Hufnagel et al., 2020; Kocira et al., 2020; Dingha et al., 2021; Rodríguez et al., 2022). In addition, legumes contribute to improved soil structure by increasing soil organic matter and aggregation, which enhances water retention, aeration, and root development. These changes foster a more favorable environment for beneficial soil microorganisms, including those involved in nutrient cycling and plant defense induction (Drinkwater et al., 1998; Blanco-Canqui et al., 2015).

Enhanced microbial activity associated with legume cover crops can also contribute to pest suppression through mechanisms such as competition, antibiosis, and the induction of systemic resistance in plants. For example, rhizosphere-associated microbes stimulated by legume residues can improve plant immune responses, thereby reducing susceptibility to insect pests and soil-borne pathogens (Berendsen et al., 2012; Pineda et al., 2017). Consequently, crops grown following legume cover crops are often more resilient and less prone to pest outbreaks compared to those grown in degraded or nutrient-poor soils (Rodríguez et al., 2022).

Legume cover crops also play a critical role in weed suppression, which has important implications for pest management. By forming dense canopies and producing substantial biomass, many legume species effectively compete with weeds for light, water, and nutrients. This reduces weed emergence and growth, thereby limiting the availability of alternative host plants that can harbor insect pests and serve as reservoirs for their populations (Dingha et al., 2021; Teasdale, 1996). In addition, some legume species release allelopathic compounds that further inhibit weed germination and development (Cherr et al., 2006). As a result, the suppression of weed communities contributes indirectly to disrupting pest life cycles and lowering infestation levels (Gazoulis et al., 2025).

Relevance to Egyptian Agroecosystems

In Egyptian agroecosystems, the integration of leguminous cover crops is particularly relevant due to the predominance of arid and semi-arid conditions, low soil organic matter, and increasing pressure on land and water resources. The use of legumes such as faba bean, short season Egyptian clover (Fahl cv.), and cowpea significantly enhances soil fertility through biological nitrogen fixation, reducing dependence on synthetic fertilizers that are often costly and inefficient under local conditions (Ouda et al., 2020; Zohry et al., 2020; El-Ramady et al., 2020). Moreover, improved soil structure and organic matter content contribute to better water retention and resilience against drought—key challenges in Egyptian agriculture (FAO, 2022). Legume cover crops also suppress problematic weed species common in Egypt, such as *Phalaris minor* and *Chenopodium album*, thereby limiting alternative hosts for insect pests and reducing pest carryover between cropping seasons (Gad et al., 2018). Additionally, the inclusion of legumes in rotations or intercropping systems enhances on-farm biodiversity and supports natural enemies of pests, contributing to more sustainable and ecologically based pest management strategies (Elsalahy et al., 2019). These combined benefits make legume cover crops a valuable tool for advancing climate-smart and resource-efficient agricultural systems in Egypt.

Conclusion

Legume-based cropping systems represent a powerful ecological approach to achieving sustainable pest and weed management in modern agriculture. By enhancing biodiversity, improving habitat structure, and supporting natural enemies, legumes play a central role in strengthening biological control processes. Intercropping systems such as faba bean and cowpea with cereals effectively reduce insect pest incidence and weed pressure through complementary resource use, physical interference, and ecological interactions. Furthermore, legume-based rotations and cover crops contribute to improved soil health, increased nutrient availability, and greater system resilience. Evidence from Egyptian agroecosystems confirms the effectiveness of these systems in reducing pest populations, suppressing weeds, and improving overall crop performance under challenging environmental conditions. Collectively, these findings underscore the importance of integrating legumes into cropping systems as a key strategy for ecological intensification, reduced chemical dependency, and the development of climate-smart agriculture.

References

Abd-Elsamed, A. A., Hashem, M. S., & Salah, A. A. A. (2011). Effect of some intercropping systems and potassium fertilization rate for maize with cowpea crops on the population density of certain pests and their natural enemies in Shrakia governorate, Egypt. *Journal of Plant Protection and Pathology, 2*(2), 153–167.

Abdel-Wahab, T. I., Abdel-Wahab, S. I., Shams, A. S., Taha, A. M., Saied, S. M., Adel, M. M., et al. (2021). Study of productivity and economic evaluation of intercropping cowpea with maize under three irrigation water levels and their response to insect infestation and viral infection. *Plant Archives, 21*(Supplement 1), 996–1019.

Abdulkareem, B. M., Mokhtassi-Bidgoli, A., Ayyari, M., Keshtkar, E., & Eyni-Nargeseh, H. (2024). Enhancing cotton sustainability: Multi-factorial intercropping, irrigation, and weed effects on productivity, quality and physiology. *Heliyon, 10*, 27135.

Abou-Keriasha, M. A., Ibrahim, S. I., & Mohamadain, E. E. A. (2011). Effect of cowpea intercropping date in maize and sorghum fields on productivity and infestation weed. *Egyptian Journal of Agronomy, 33*(1), 35–49.

Abou-Keriasha, M. A., Mohamed, W. M. A., Eiasa, N. M., & Kamel, A. S. (2012). Intensive crop rotations to improve agricultural production in Middle Egypt. *Egyptian Journal of Agricultural Research, 90*(4), 427–443.

Abou-Keriasha, M. A., Eisa, N. M. A., & Lamlom, M. M. (2013). Benefits of legume crops in rotation and intercropping for increased production and land use. *Egyptian Journal of Agronomy, 35*(2), 183–197.

Al-Musawi, Z. K., Vona, V., & Kulmány, I. M. (2025). Utilizing different crop rotation systems for agricultural and environmental sustainability: A review. *Agronomy, 15*, 1966. https://doi.org/10.3390/agronomy15081966

Bach, C. E. (1981). Host plant growth form and diversity: Effects on abundance and feeding preference of a specialist herbivore, Acalymma vittata (Coleoptera: Chrysomelidae). *Oecologia, 50*, 370–375.

Bedoussac, L., Journet, E. P., Hauggaard-Nielsen, H., et al. (2015). Ecological principles underlying the increase of productivity achieved by cereal–grain legume intercrops. *Agronomy for Sustainable Development, 35*(3), 911–935. https://doi.org/10.1007/s13593-014-0277-7

Berendsen, R. L., Pieterse, C. M. J., & Bakker, P. A. H. M. (2012). The rhizosphere microbiome and plant health. *Trends in Plant Science, 17*(8), 478–486.

Blanco-Canqui, H., Shaver, T. M., Lindquist, J. L., Shapiro, C. A., Elmore, R. W., Francis, C. A., & Hergert, G. W. (2015). Cover crops and ecosystem services: Insights from studies in temperate soils. *Agronomy Journal, 107*, 2449–2474.

Bowles, T. M., Mooshammer, M., Socolar, Y., Calderón, F., Cavigelli, M. A., et al. (2020). Long-term evidence shows that crop-rotation diversification increases agricultural resilience to adverse growing conditions in North America. *One Earth, 2*(3), 284–293. https://doi.org/10.1016/j.oneear.2020.02.007

Chau, N. N. B., Kieu, N. T. P., Dung, N. V. T., Quoc, N. B., & Phuong, T. K. (2019). Effects of floral resources on the longevity and parasitism of *Cotesia Vestalis* Haliday (Hymenoptera: Braconidae) on *Plutella Xylostella* (L.) (Lepidoptera: Plutellidae) in Vietnam. *Heliyon, 5*(8), e02258. https://doi.org/10.1016/j.heliyon.2019.e02258

Cherr, C. M., Scholberg, J. M. S., & McSorley, R. (2006). Green manure approaches to crop production: A synthesis. *Agronomy Journal, 98*, 302–319.

Cole, L. J., Baddeley, J. A., Robertson, D., et al. (2022). Supporting wild pollinators in agricultural landscapes through targeted legume mixtures. *Agriculture, Ecosystems & Environment, 323*, 107648.

Dhima, K., Vasilakoglou, I., Gatsis, T., & Gougoulias, N. (2018). Faba bean-barley intercrops for high productivity and corn poppy suppression. *Experimental Agriculture, 54*(2), 163–180. https://doi.org/10.1017/S0014479716000132

Dingha, B. N., Omaliko, P. C., Amoah, B. A., Jackai, L. E., & Shrestha, D. (2021). Evaluation of cowpea (*Vigna unguiculata*) in an intercropping system as pollinator enhancer for increased crop yield. *Sustainability, 13*, 9612. https://doi.org/10.3390/su13179612

Drinkwater, L. E., Wagoner, P., & Sarrantonio, M. (1998). Legume-based cropping systems have reduced carbon and nitrogen losses. *Nature, 396*(19), 262–265.

Ediagbonya, T. F., Areo, I. O., Mupenzi, C., et al. (2025). Reduced pesticide dependency through crop management. *Discovery and Applied Sciences, 7*, 776. https://doi.org/10.1007/s42452-025-07248-y

El-Mehy, A. A., El-Gendy, H. A., Aioub, A. A. A., Mahmoud, S. F., Abdel-Gawad, S., et al. (2022). Response of faba bean to intercropping, biological and chemical control against broomrape and root rot diseases. *Saudi Journal of Biological Sciences., 29*(5), 3482–3493. https://doi.org/10.1016/j.sjbs.2022.02.032

El-Ramady, H., Abdalla, N., Kovacs, S., Szabolcsy, E. D., Bákonyi, N., et al. (2020). Sustainable biorefinery of alfalfa (Medicago sativa L.): A review. *Egyptian Journal of Botany, 60*(3), 621–639.

Elsalahy, H., Döring, T., Bellingrath-Kimura, S., & Arends, D. (2019). Weed suppression in only-legume cover crop mixtures. *Agronomy, 9*, 648. https://doi.org/10.3390/agronomy9100648

Eskandari, H., & Kazemi, K. (2011). Weed control in maize-cowpea intercropping system related to environmental resources consumption. *Not Sci Biol, 3*(1), 57–60.

FAO. (2022). *Towards climate-smart agriculture in Egypt*. Food and Agriculture Organization of the United Nations.

Finch, S., & Collier, R. H. (2000). Host-plant selection by insects—A theory based on 'appropriate/inappropriate landings' by pest insects of cruciferous plants. *Entomologia Experimentalis et Applicata, 96*, 91–102.

Foyer, C. H., Lam, H. M., Nguyen, H. T., et al. (2016). Neglecting legumes has compromised human health and sustainable food production. *Nature Plants, 2*, 1–10.

Gad, A. A., Ali, M. A., Abd El-Lattief, E. A., & Mekky, M. S. (2018). Study impact of integration between cover crop and weed control treatments on weeds and improved sugarcane productivity. *Assiut Journal of Agricultural Sciences, 49*(4), 32–43.

Gazoulis, K. M. I., Kontogeorgou, D. M. V., Kanatas, P., & Travlos, I. (2025). Enhancing ecosystem services in agriculture: The special role of legume intercropping. *Frontiers in Sustainable Food Systems, 9*, 1547879. https://doi.org/10.3389/fsufs.2025.1547879

Gitari, H. I., Nyawade, S. O., Kamau, S., Karanja, N. N., Gachene, C. K. K., Raza, M. A., Maitra, S., & Schulte-Geldermann, E. (2020). Revisiting intercropping indices with respect to potato -legume intercropping systems. *Field Crops Research, 258*, 107957.

Gu, C., Bastiaans, L., Anten, N. P. R., Makowski, D., & Vander Werf, W. (2021). Annual intercropping suppresses weeds: A meta-analysis. *Agriculture, Ecosystems & Environment, 322*, 107658.

Guinet, M., Adeux, G., Cordeau, S., Courson, E., Nandillon, R., Zhang, Y., & Munier-Jolain, N. (2023). Fostering temporal crop diversification to reduce pesticide use. *Nature Communications, 14*(1), 7416. https://doi.org/10.1038/s41467-023-43234-x. PMID: 37973850; PMCID: PMC10654721.

Gurr, G. M., Wratten, S. D., Landis, D. A., & You, M. (2017). Habitat management to suppress Pest populations: Progress and prospects. *Annual Review of Entomology, 62*, 91–109.

Hamd-Alla, W. A., Shalaby, E. M., Dawood, R. A., & Zohry, A. A. (2014). Effect of cowpea (Vigan sinensis L.) with maize (Zea mays L.) intercropping on yield and its components. *Inter. Scho. Sci. Res. Innov., 8*(11), 1170–1176.

Hegazi, E., Abou Zeid, M. A. A., Hasaneen, M. A.-E., & al Abu Shall, A. e. (2024). Effect of intercropping by flax, radish and fenugreek on faba bean, *Vicia faba* L., production and reduction of *Orobanche crenata* forsk seed bank. *Agriculture, Forestry and Fisheries, 13*(2), 52–59. https://doi.org/10.11648/j.aff.20241302.15

Hufnagel, J., Reckling, M., & Ewert, F. (2020). Diverse approaches to crop diversification in agricultural research. A review. *Agronomy for Sustainable Development, 40*. https://doi.org/10.1007/s13593-020-00617-4

Jamshidi, K., Yousefi, A. Z., & Oveisi, M. (2013). Effect of cowpea (Vigna unguiculata) intercropping on weed biomass and maize (Zea mays) yield. *New Zealand Journal of Crop and Horticultural Science, 41*(4), 180–188. https://doi.org/10.1080/01140671.2013.807853

Jensen, E. S., Carlsson, G., Hauggaard-Nielsen, H., & Eriksen, J. (2020). Intercropping of grain legumes and cereals improves the use of soil N resources and reduces the requirement for synthetic fertilizer N: A global-scale analysis. *Agronomy for Sustainable Development, 40*(1), 5. https://doi.org/10.1007/s13593-020-0607-x. ⟨hal03141077⟩.

Katsaruware, R. D., & Manyanhaire, I. O. (2009). Maize-cowpea intercropping and weed suppression in leaf stripped and detasselled maize in Zimbabwe. *Electronic Journal of Environmental, Agricultural and Food Chemistry, 8*(11), 1218–1226.

Kazemi, K., Eskandari, H., & Mosavian S., N. (2025). More environmental resources consumption in low-input intercropped maize-cowpea leads to the suppression of weed growth. *Journal of Crop Protection, 14*(2), 133–143. https://doi.org/10.48311/jcp.2025.76211.0

Kemper, R., Rinke, N., Gerhards, R., & Böhm, H. (2020). Weed suppression and crop yield performance in sole and intercrops of common vetch and spring wheat depending on seed density ratio in organic farming. *Journal für Kulturpflanzen, 72*, 12–24.

Kirsch, F., Hass, A. L., Link, W., et al. (2023). Intercrops as foraging habitats for bees: Bees do not prefer sole legume crops over legume-cereal mixtures. *Agriculture, Ecosystems & Environment, 343*, 108268.

Kocira, A., Staniak, M., Tomaszewska, M., Kornas, R., Cymerman, J., Panasiewicz, K., & Lipińska, H. (2020). Legume cover crops as one of the elements of strategic weed management and soil quality improvement. A Review. *Agriculture, 10*, 394. https://doi.org/10.3390/agriculture10090394

Köpke, U., & Nemecek, T. (2010). Ecological services of faba bean. *Field Crops Research, 115*, 217–233.

Kumar, S., Bhowmick, M., & Ray, P. (2021). Weeds as alternate and alternative hosts of crop pests. *Indian Journal of Weed Science, 53*, 14–29.

Lebreton, P., Bedoussac, L., Bonnet, C., Journet, E.-B., Justes, E., & Colbach, N. (2024). Optimal species proportions, traits and sowing patterns for agroecological weed management in legume–cereal intercrops. *European Journal of Agronomy, 159*, 127266. https://doi.org/10.1016/j.eja.2024.127266

Leoni, F., Lazzaro, M., Ruggeri, M., et al. (2022). Relay intercropping can efficiently support weed management in cereal-based cropping systems when appropriate legume species are chosen. *Agronomy for Sustainable Development, 42*, 75. https://doi.org/10.1007/s13593-022-00787-3

Letourneau, D. K., Armbrecht, I., Rivera, B. S., et al. (2011). Does plant diversity benefit agroecosystems? A synthetic review. *Ecological Applications, 21*, 9–21.

Li, L., Zhang, L.-Z., & Zhang, F.-Z. (2013). Crop mixtures and the mechanisms of overyielding. In S. A. Levin (Ed.), *Encyclopedia of biodiversity, 2nd edition, Vol. 2* (pp. 382–395). Academic Press.

Lithourgidis, A. S., Dordas, C. A., Damalas, C. A., & Vlachostergios, D. N. (2011). Annual intercrops: An alternative pathway for sustainable agriculture. *Australian Journal of Crop Science, 5*(4), 396–410.

Lu, Z. X., Zhu, P. Y., Gurr, G. M., Zheng, X. S., Read, D. M. Y., Heong, K. L., Yang, Y. J., & Xu, H. X. (2014). Mechanisms for flowering plants to benefit arthropod natural enemies of insect pests: Prospects for enhanced use in agriculture. *Insect Science, 21*(1), 1–12. https://doi.org/10.1111/1744-7917.12000

Mohamed, H. O., Dahi, H. F., & Mahmoud, A. H. (2023). Diversity of harmful and beneficial insects on faba bean, *Vicia faba* L. in Assiut Governorate, Northern Upper Egypt. *Academia Biology, 1*. https://doi.org/10.20935/AcadBiol6154

Mukanga, M., Machuku, O., Lwinya, K., Lupapula, M., Matimelo, M., & Chilipa, L. (2024). Effect of intercropping maize with legumes, oilseed crops and cucurbits, and perimeter cropping on fall armyworm (Spodoptera frugiperda) Infestation in Zambia. *Journal of Agriculture and Environmental Sciences, 13*, 16–28.

Ouda, S. A., Kasem E., S., Zohry, A., & Abdel-Wahab, S. (2020). Sustainable intensive cropping to reduce irrigation-induced erosion: II. Changing cropping sequence under sprinkler irrigation practice. *Moroccan Journal of Agricultural Science., 1*(1), 26–33.

Ouyang, F., Su, W., Zhang, Y., Liu, X., Su, J., Zhang, Q., Men, X., Ju, Q., & Ge, F. (2020). Ecological control service of the predatory natural enemy and its maintaining mechanism in rotation-intercropping ecosystem via wheat-maize-cotton, agriculture. *Agriculture, Ecosystems & Environment, 301*, 107024. https://doi.org/10.1016/j.agee.2020.107024

Pierre JF, Latournerie-Moreno L, Garruña-Hernández R, Jacobsen KL, Laboski CAM, Salazar-Barrientos LdL, Ruiz-Sánchez E. (2021). Farmer perceptions of adopting novel legumes in traditional maize-based farming systems in the Yucatan peninsula. *Sustainability 13*(20):11503 https://doi.org/10.3390/su132011503.

Pierre, J. F., Latournerie-Moreno, L., Garruña, R., Jacobsen, K. L., Laboski, C. A. M., Us-Santamaría, R., & Ruiz-Sánchez, E. (2022). Effect of maize–legume intercropping on maize physio-agronomic parameters and beneficial insect abundance. *Sustainability, 14*, 12385. https://doi.org/10.3390/su141912385

Pierre, J. F., Jacobsen, K. L., Latournerie-Moreno, L., Torres-Cab, W. J., Chan-Canché, R., & Ruiz-Sánchez, E. (2023). A review of the impact of maize-legume intercrops on the diversity and abundance of entomophagous and phytophagous insects. *PeerJ, 11*, e15640. https://doi.org/10.7717/peerj.15640

Pineda, A., Kaplan, I., & Bezemer, T. M. (2017). Steering soil microbiomes to suppress aboveground insect pests. *Trends in Plant Science, 22*(9), 770–778.

Plaza, M. E. H., Bastida, F., Gibson, D. J., Barro, F., Giménez, M. J., Pallavicini, Y., Izquierdo, J., & González-Andújar, J. L. (2022). Grain quality as influenced by the structural properties of weed communities in Mediterranean wheat crops. *Agronomy, 13*, 49.

Rahimi, I. M., Amanullah, M., Ananthi, T., & Mariappan, G. (2019). Influence of intercropping and weed management practices on weed parameters and yield of maize. *International Journal of Current Microbiology and Applied Sciences, 8*(4), 2167–2172. https://doi.org/10.20546/ijcmas.2017.604.175

Rao, M. S., Prasad, T. V., Gayatri, D. L. A., Rao, C. A. R., Srinivas, K., et al. (2025). Impact of intercropping and low-external input integrated pest management on Spodoptera frugiperda (J. E. Smith) in maize. *Field Crops Research, 326*, 109868. https://doi.org/10.1016/j.fcr.2025.109868

Rodríguez, B. C., Zuazo, V. H. D., Rodríguez, M. S., García-Tejero, I. F., Ruiz, B. G., et al. (2022). Legumes protect the soil erosion and ecosystem services. In *Advances in legumes for sustainable intensification* (pp. 247–266). Academic Press. https://doi.org/10.1016/B978-0-323-85797-0.00010-0

Root, R. B. (1973). Organization of plant–arthropod association in simple and diverse habitats: The fauna of collards (Brassica oleraceae). *Ecological Monographs, 43*, 95–124.

Rosa-Schleich, J., Loos, J., Mußhoff, O., et al. (2019). Ecological economic trade-offs of diversified farming systems–a review. *Ecological Economics, 160*, 251–263.

Ruhanen, H. M., Räty, E., Mäkinen, J., et al. (2024). Effect of faba bean nectar on longevity and fecundity of *Plutella xylostella* and its parasitoid *Cotesia vestalis*. *Arthropod-Plant Interactions, 18*, 455–468. https://doi.org/10.1007/s11829-024-10066-5

Sammama, H., Alfeddy, M. N., Hsissou, D., & El Kaoua, M. (2023). Potential effect of intercropping in the control of weeds, diseases, and pests in a wheat-faba bean system. *Acta Agriculturae Slovenica, 119*(1), 1–11.

Saudy, H. S. (2015). Maize-cowpea intercropping as an ecological approach for nitrogen-use rationalization and weed suppression. *Archives of Agronomy and Soil Science, 61*(1), 1–14. https://doi.org/10.1080/03650340.2014.920499

Shah, S. N., Shroff, J. C., Patel, R. H., & Usadadiya, V. P. (2011). Influence of intercropping and weed management practices on weed and yields of maize. *International Journal of Science and Nature., 2*(1), 47–45.

Sharma, R. C., & Banik, P. (2013). Baby corn-legumes intercropping system: II weed dynamics and community structure. *NJAS - Wageningen Journal of Life Sciences, 67*, 11–18. https://doi.org/10.1016/j.njas.2013.08.001

Sharma, G., Shrestha, S., Kunwar, S., & Tseng, T. (2021). Crop diversification for improved weed management: A review. *Agriculture, 11*, 461.

Singh, N., & Kumar, R. (2023). Legume intercropping for weed control efficiency in kharif maize (Zea mays l.) under Amritsar conditions. *Agricultural Science Digest*. https://doi.org/10.18805/ag.D-5656

Singh, H., Reager, M. L., & Singh, S. P. (2017). Effects of weed control measures and intercropping on weed biomass. In *Biennial conference of the Indian Society of Weed Science on "doubling farmers' income by 2022: The role of weed science"*. MPUA&T. during 1-3 March, pag74.

Snapp, S. S., Cox, C. M., & Peter, B. G. (2019). Multipurpose legumes for smallholders in sub-Saharan Africa: Identification of promising 'scale out' options. *Global Food Security, 23*, 22–32. https://doi.org/10.1016/j.gfs.2019.03.002

Soujanya, P. L., VaniSree, K., Giri, G. S., Mahadik, S., Jat, S. L., Sekhar, J. C., et al. (2024). Intercropping in maize reduces fall armyworm Spodoptera frugiperda (J. E. Smith) infestation, supports natural enemies, and enhances yield. *Agriculture, Ecosystems & Environment, 373*, 109130. https://doi.org/10.1016/j.agee.2024.109130

Stenberg, J. A., Sundh, I., Becher, P. G., et al. (2021). When is it biological control? A framework of definitions, mechanisms, and classifications. *Journal of Pest Science, 94*, 665–676. https://doi.org/10.1007/s10340-021-01354-7

Suhi, A., Mia, S., Khanam, S., Hasan Mithu, M., Uddin, M. K., Muktadir, M. A., Ahmed, S., & Jindo, K. (2022). How does maize-cowpea intercropping maximize land use and economic return? A field trial in Bangladesh. *Land, 11*, 581. https://doi.org/10.3390/land11040581

Suso, M. J., Bebeli, P. J., Christmann, S., Mateus, C., Negri, V., Pinheiro de Carvalho, M. A. A., Torricelli, R., & Veloso, M. M. (2016). Enhancing Legume Ecosystem Services through an Understanding of Plant–Pollinator Interplay. *Frontiers in Plant Science, 7*, 333. https://doi.org/10.3389/fpls.2016.00333

Takim, F. O. (2012). Advantages of maize-cowpea intercropping over sole cropping through competition indices. *Journal of Agriculture and Biodiversity Research, 1*(4), 53–59. file:///C:/Users/smart/Downloads/TakimJABR2012.pdf.

Teasdale, J. R. (1996). Contribution of cover crops to weed management. *Journal of Production Agriculture, 9*, 475–479.

Teasdale, J. R. (2017). The use of rotations and cover crops to manage weeds. In R. L. Zimdahl (Ed.), *Integrated weed management for sustainable agriculture* (pp. 227–260). Burleigh Dodds.

Vamdermeer, J. (1989). *The ecology of intercropping* (p. 237). Cambridge University Press.

Verret, V., Gardarin, A., Pelzer, E., Médiène, S., Makowski, D., & Valantin-Morison, M. (2017). Can legume companion plants control weeds without decreasing crop yield? A meta-analysis. *Field Crops Research, 204*, 158–168. https://doi.org/10.1016/j.fcr.2017.01.010

Villegas-Fernández, Á. M., Amarna, A. A., Moral, J., & Rubiales, D. (2024). Intercropping as a strategy for weed management in Faba Bean (*Vicia faba* L.). *Agronomy, 14*, 1170. https://doi.org/10.3390/agronomy14061170

Vrignon-Brenas, S., Celette, F., Piquet-Pissaloux, A., Jeuffroy, M.-H., & David, C. (2016). Early assessment of ecological services provided by forage legumes in relay intercropping. *European Journal of Agronomy, 75*, 89–98. https://doi.org/10.1016/j.eja.2016.01.011

Weiner, M., Moakes, S., Raya-Sereno, M. D., & Cooper, J. (2024). Legume-based crop rotations as a strategy to mitigate fluctuations in fertilizer prices? A case study on bread wheat genotypes in northern Spain using life cycle and economic assessment. *European Journal of Agronomy, 159*, 127267. https://doi.org/10.1016/j.eja.2024.127267

Weisberger, D., Nichols, V., & Liebman, M. (2019). Does diversifying crop rotations suppress weeds? A meta-analysis. *PLoS One, 14*(7), e0219847. https://doi.org/10.1371/journal.pone.0219847. PMID: 31318949; PMCID: PMC6638938.

Wyckhuys, K. A. G., Gonzalez-Chang, M., Lavandero, B. Y. G. C., & Hadi, B. A. R. (2023). Legume integration for biological control deserves a firmer scientific base. *Outlook on Agriculture*, 1–4. https://doi.org/10.1177/00307270231194086. journals.sagepub.com/home/oag.

Zacarias, D. A. (2020). Global bioclimatic suitability for the fall armyworm, Spodoptera frugiperda (Lepidoptera: Noctuidae), and potential co-occurrence with major host crops under climate change scenarios. *Climatic Change.*, Springer, *161*(4), 555–566.

Zhang, D. B., Yao, Z. Y., & al Chen, J. e. (2019). Improving soil aggregation, aggregate-associated C and N, and enzyme activities by green manure crops in the loess plateau of China. *European Journal of Soil Science, 70*, 1267–1279. https://doi.org/10.1111/ejss.12843

Zohry, A. A. (2005). Effect of preceding winter crops and intercropping on yield, yield components and associated weeds in maize. *Annals of Agricultural Science, Moshtohor, 43*(1), 139–148.

Zohry, A., Ouda, S., & Abdel-Wahab, T. (2020). Sustainable intensive cropping to reduce irrigation-induced erosion: I. Intercropping systems under surface irrigation practice. *Moroccan Journal of Agricultural Science, 1*(2), 63–71.

Zou, Y., Liu, Z., Chen, Y., Wang, Y., & Feng, S. (2024). Crop rotation and diversification in China: Enhancing sustainable agriculture and resilience. *Agriculture, 14*, 1465. https://doi.org/10.3390/agriculture14091465

Chapter 8
The Agroecosystem Services of Cowpea Intercropping System with Maize in the Context of Egypt

Abstract This chapter evaluated the agroecosystem services and productivity of a cowpea–maize intercropping system in Middle Egypt under varying irrigation levels and cowpea planting dates. A two-year field experiment was conducted using three irrigation regimes (120%, 100%, and 80% ETc) and three cowpea planting dates (2 weeks before maize, simultaneous with maize, and 2 weeks after maize). Intercropping improved maize resilience, enhanced land and water productivity, increased soil organic matter, and elevated available soil nitrogen and phosphorus compared with monoculture maize. The highest land equivalent ratio occurred when cowpea was sown simultaneously with maize, while water equivalent ratio was maximized when cowpea was sown after maize. Intercropping reduced weed biomass and improved microclimate buffering under water stress. Economic returns and monetary advantage indices were consistently higher in intercropping systems than in sole maize. The calibrated and validated SALTMED model successfully simulated maize performance under present conditions and was used to project future yields under SSP5-8.5 climate scenario. Future projections indicated substantial yield declines for sole maize under climate change, whereas intercropping with cowpea mitigated productivity losses—particularly when cowpea was planted after maize. Overall, cowpea–maize intercropping demonstrated strong potential to enhance agroecosystem sustainability, climate resilience, and resource-use efficiency in arid Egyptian environments.

Keywords Land equivalent ratio · Water equivalent ratio · Monetary advantage index · Soil organic matter · Soil nitrogen and phosphorus · Weed biomass · SALTMED model · SSP5-8.5 future climate scenario

Introduction

Legumes are essential for rejuvenating soil health and present a viable approach for agricultural and ecological sustainability (Kumawata et al., 2022). Legume crop residues enhance microbial population and functionality by supplying energy and

A. E. Zohry, S. A. H. Ouda, *The Agroecosystem Services of Leguminous Crops*, https://doi.org/10.1007/978-3-032-32461-0_8

creating an optimal soil environment, contributing substantial organic matter, mitigating soil and wind erosion, and improving soil structure and water retention capacity (Lal, 2011). Furthermore, legumes play a crucial role in the agroecosystem by turning atmospheric inert nitrogen into a plant-available form through biological nitrogen fixation in combination with symbiotic bacteria, hence enhancing soil health (Lal, 2017) (Fig. 8.1).

Intercropping is a significant agronomic technique that entails cultivating two or more crops concurrently on the same land area (Kumar et al., 2020). Intercropping often involves a primary crop, distinguished by its economic advantage or nutritional worth, alongside one or more supplementary crops (Dwivedi et al., 2015). Intercropping is an effective and economical production system that enhances yield per unit area and time, improves resource use efficiency, and increases the economic viability of growers. Additionally, it acts as a safeguard against total crop failure and contributes to food security (Bhatti et al., 2013).

Legume-based intercropping systems are essential for providing various ecosystem services, including the production of protein-rich food for humans and livestock (Tharanathan & Mahadevamma, 2003). They are recommended for low-input production systems due to their capacity to enhance soil fertility through biological nitrogen fixation (Jangir et al., 2017), increase soil organic content (Mitran et al., 2018), and reduce greenhouse gas emissions (Kumar et al., 2018). Legume-based intercropping systems achieve crop protection by diversifying cropping systems (Nemecek et al., 2008). Furthermore, legume intercropping mitigates the adverse effects of weeds on crop yield by inhibiting their development and biomass (Jensen et al., 2006). Consequently, legume-based intercropping systems are ecologically, physiologically, and socioeconomically advantageous compared to intense sole cropping (He et al., 2012).

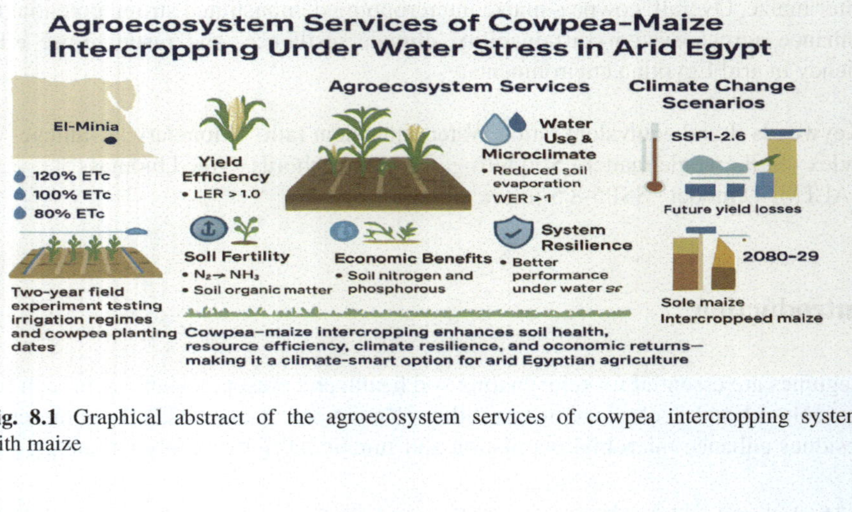

Fig. 8.1 Graphical abstract of the agroecosystem services of cowpea intercropping system with maize

Maize ranks among the most significant crops globally, with production exceeding 1.2 billion metric tons (FAOSTAT, 2025). It is essential to world agriculture, functioning as a fundamental food source for people and a vital component in animal feed (Ouda et al., 2020). In addition to direct consumption, maize serves as a crucial raw material for other sectors, such as the manufacturing of corn starch, corn syrup, and ethanol. In Egypt, maize is a vital summer cereal crop, encompassing the most extensive planted area. A significant disparity exists between domestic production and consumption, as the nation's maize self-sufficiency ratio is at 47% (Ouda et al., 2020). Research initiatives in Egypt have concentrated on enhancing maize yield via effective irrigation techniques, improved fertilization, the creation of novel inbreds and hybrids, and the strategic application of organic and biofertilizers (Mansour, 2025). Improved water management coupled with appropriate nitrogen fertilizer has demonstrated a substantial improvement in yields and resource-use efficiency (Gomaa et al., 2021). Moreover, an efficient strategy to augment maize productivity is to intercrop cowpea, which improves maize yield and enhances water usage efficiency.

Cowpea is an annual leguminous forage crop recognized for its moderate drought tolerance and farmed for its nutritious cattle feed. It can also function as an efficient cover crop that reduces soil erosion and lessens soil evaporation (Singh et al., 2003). Furthermore, cowpea is characterized by its shade tolerance, rendering it appropriate for intercropping with various cereals and root crops (Boukar et al., 2013). Cowpea, known for its fast proliferation, can be farmed as either a monocrop or an intercrop, and possesses the unique ability to fix nitrogen even in severely deficient soils (Singh, 2014). The biological nitrogen fixation generated by cowpea varies from 36 to 75 kg/ha (Vasconcelos et al., 2020).

Intercropping cowpea with maize is an efficient and sustainable agricultural strategy that provides numerous benefits to the agroecosystem. Arshad (2025) documented an enhancement in soil carbon sequestration when cowpea was intercropped with maize. Roohi et al. (2022) contend that intercropping cowpea with maize improves carbon sequestration compared to monocultures of either crop. Furthermore, Ghosh et al. (2007) asserted that cowpea intercropped with maize positively influenced the yield of the subsequent wheat crop.

It has been stated that cowpea intercropping with maize substantially modifies the microclimate within the canopy, chiefly by altering light intensity, temperature, and soil moisture (Ghanbari et al., 2010). The benefits of cowpea intercropping with maize may progress from improved moisture retention in intercrops compared to sole crops, as indicated by Dube et al. (2014). The cowpea improved early ground cover, decreasing soil evaporation, while maize moderated canopy temperature, thereby enhancing maize's resilience under microclimatic stress (Haruna et al., 2018). According to Ghosh et al. (2007), intercropping cowpea with maize positively affected the yield of the subsequent crop due to its residual impact on soil fertility. Furthermore, intercropping cowpea with maize enhanced nutrient recycling and facilitated biological nitrogen fixing (Dimande et al., 2024). The availability of phosphorus in the rhizosphere was enhanced in a cowpea intercropping system, accompanied by notable pH alterations (Latati et al., 2014). Ferguson et al. (2013) demonstrated that legumes are more adept at extracting calcium and magnesium from the soil than cereals, substituting these elements with hydrogen, which leads to the removal of OH^- ions and an increase in H^+, consequently lowering soil pH. The contribution of cowpea to soil conservation was noted, as it establishes ground cover early in the season, in contrast to maize (Haruna et al., 2018). Additionally, intercropping cowpea with maize has proven effective in mitigating soil erosion (Akchaya et al., 2025). Tamta et al. (2019) reported that intercropping cowpeas with maize (two rows of maize for every one row of cowpea) diminished runoff by 10% and soil erosion by 28% in comparison to maize monoculture. Moreover, the cowpea intercropping system with maize diminishes related weeds (Zohry, 2005) and insects (Hamd-Alla et al., 2014) in comparison to maize monoculture.

Prior investigations in Egypt regarding the cowpea–maize intercropping system primarily concentrated on enhancing maize yield and land use efficiency by examining various intercropping configurations of cowpea (Elsaid et al., 2019; El-Ghobashy et al., 2018; El-Shamy et al., 2022), fertilizer application (El-Ghobashy et al., 2020; Shehata & Hamd-Alla, 2023), and the interaction between intercropping patterns and fertilizer application (Salama et al., 2022). Only two studies examined the impact of cowpea planting dates on maize under intercropping (Abou-Keriasha et al., 2011; El-Shamy et al., 2021), while one study explored the influence of water stress on the cowpea–maize intercropping system (Taha & Kasem, 2022). However, the impact of varying planting dates of cowpea in an intercropping system with maize, alongside water stress, on the yield of both crops and the agroecosystem services of cowpea–maize intercropping system in Egypt has not been investigated.

This chapter evaluated the agroecosystem services and productivity of a cowpea–maize intercropping system in Middle Egypt under varying irrigation levels and cowpea planting dates. A comparative study was conducted to assess the effectiveness of cowpea in an intercropping system with maize regarding agroecosystem services. The study evaluated the impact of varying cowpea planting dates and imposed water stress on the yields of both crops, as well as water and land use efficiencies and farmer benefits.

Cowpea Intercropping Systems with Maize: The Field Experiment

A two-year field experiment was conducted in El-Minia Governorate (latitude 28.05°, longitude 30.44°, elevation 40.0 m) in Middle Egypt during the 2022 and 2023 growing seasons, utilizing sandy soil under a drip irrigation system. The experimental area has an arid climate marked by cold winters and sweltering, arid summers (Masoued, 2017). Data on average daily air temperature, relative humidity, and wind speed for the research seasons were obtained from the weather station located 320 m from the experimental site.

The soil at the experimental site is classified as sandy, with 0.31% organic matter. The physical and chemical properties of the soil were evaluated according to the methods established by Klute (1986) and Tan (1996) (Tables 8.1 and 8.2). The irrigation water was obtained from groundwater, with an electrical conductivity of 0.92 dS/m and a pH of 7.53.

The experimental design employed a split-plot design with three replications. The treatments comprised the interaction of three irrigation levels with three planting dates of cowpea (*Vigna unguiculata* (L.) Walp) intercropped with maize (*Zea mays* L.), in addition to sole maize and sole cowpea cultivations. The irrigation treatments included the requisite irrigation amount (RI), set at 120% of crop evapotranspiration (ETc) as the control treatment, along with two water stress treatments: WS1, representing 100% ETc; and WS2, representing 80% ETc. The three cowpea planting schedules were: cowpea sown 2 weeks prior to maize culture (PD1), cowpea sown concurrently with maize cultivation (PD2), and cowpea sown 2 weeks subsequent to maize cultivation (PD3). The irrigation treatments were allocated to the main plots, and the three intercropping dates were designated to the sub-plots. The plot area measured 21.0 m^2 (3.0 × 7.0 m) and consisted of 10 ridges, each 0.70 m in width. The prior crop was wheat in both growing seasons, sown in November and harvested in late April.

The Sole and Intercropped Maize and Cowpea

Maize seeds were planted on one side of the ridge, with a spacing of 25 cm between hills (one plant per hill), adhering to the full required planting density for both monoculture and intercropped maize (57,600 plants per hectare, 100% recommended density). The maize hybrid TWC310 was sown on May 15 and May 21 during the first and second growing seasons, respectively, either as a monoculture or within an intercropping system. The harvest occurred on September 1 and 7 in the first and second growing seasons, respectively. Calcium super phosphate (15.5% P_2O_5) was applied at a rate of 357.0 kg/ha, while potassium sulfate (48.0% K_2O) was applied at a rate of 119.0 kg/ha during soil preparation. Mineral nitrogen fertilizer (ammonium nitrate 33.3% N) for maize was applied at a rate of 285.6 kg/ha, distributed in five equal doses every 10 days post-planting.

The cowpea cultivar Cream was employed in monoculture and in the intercropping systems. The planting density of cowpea was at the full recommended level under both sole and intercropped cowpea (72 kg seeds/ha, 100% recommended

Table 8.1 Primary physical characteristics of soil at the experimental location prior to cultivation

Soil depth (cm)	Particle size distribution			Texture class	Bulk density (g/cm³)	Field capacity (%)	Permanent wilting point (%)	Available water (%)
	Sand (%)	Silt (%)	Clay (%)					
0–20	93.40	4.70	2.11	Sandy	1.64	12.76	3.61	9.15
20–40	95.50	3.10	1.50		1.75	11.22	2.89	8.33
40–60	96.10	2.85	0.95		1.69	7.43	2.09	5.34

Table 8.2 Chemical characteristics and accessible macronutrients of the soil at the experimental location prior to cultivation

Soil depth	pH	Ece	Soluble cations (meq/L)				Soluble anions (meq/L)			
(cm)	(1:2.5)	(dS m^{-1})	Ca^{+2}	Mg^{+2}	Na^{+}	K^{+}	CO_3^{-2}	HCO_3^{-}	Cl^{-}	SO_4^{-2}
0–20	7.61	0.690	2.42	1.08	3.12	0.32	–	2.22	3.44	1.28
20–40	7.58	0.692	2.38	1.02	3.12	0.30	–	2.08	3.48	1.26
40–60	7.49	0.686	2.32	1.01	3.10	0.30	–	2.14	3.46	1.13
Available nutrients (mg/kg)										
N				P_2O_5				K_2O		
12.14				4.61				58.21		

density). The sole cowpea was planted on both sides of the ridge, with 15 cm spacing between hills (one plant per hill). The intercropped cowpea was planted on the opposite side of maize ridge, with 15 cm spacing between hills (two plants per hill). Sole cowpea was sown on May 15 and May 21 in the first and second seasons, respectively. Two cuts were conducted for both sole and intercropped cowpea; the initial harvest occurred 60 days post-sowing, and the subsequent harvest coincided with the maize harvest date. Nitrogen fertilizer (ammonium nitrate, 33.3% N) was applied to cowpea at 48 kg/ha in one dose, 3 weeks after planting.

The maize grain yield, both from intercropping and monoculture, was determined by harvesting and weighing all plants within the experimental plot area, followed by aggregating the weights of all plots. The output from both cuts of cowpea was combined to determine the total forage yield. The biomass from all examined intercropping systems was gathered from the field post-harvest. The experiment was conducted at the same site utilized in the initial growing season throughout the subsequent growing season.

Data from each growing season were evaluated independently utilizing a split-plot design with three replicates. Statistical analysis was performed independently for maize and cowpea yields in both sole and intercropped scenarios, adhering to the methodologies outlined by Gomez and Gomez (1984). Mean comparisons were conducted utilizing the least significant difference (LSD) test at the 5% significance level, as delineated by Waller and Duncan (1969).

Characterization of the Meteorological Conditions at the Experimental Location

Figure 8.2 illustrates the mean temperature, solar radiation, and wind speed during the first growing season (3a) and the second growing season (3b). The figure indicated substantial fluctuations in these meteorological factors across the two growing seasons, with more pronounced variations observed in the second season. This suggested that it could influence the value of evapotranspiration (ETo), the primary determinant of water requirements.

Figure 8.3 illustrates a comparison of the computed monthly ETo values during the first and second growing seasons. The figure indicated elevated ETo values during the second growing season in comparison to the first growing season in June, July, and August. Consequently, it is anticipated that the values of applied irrigation water treatments will be elevated in the second growing season compared to the first, along with water consumption.

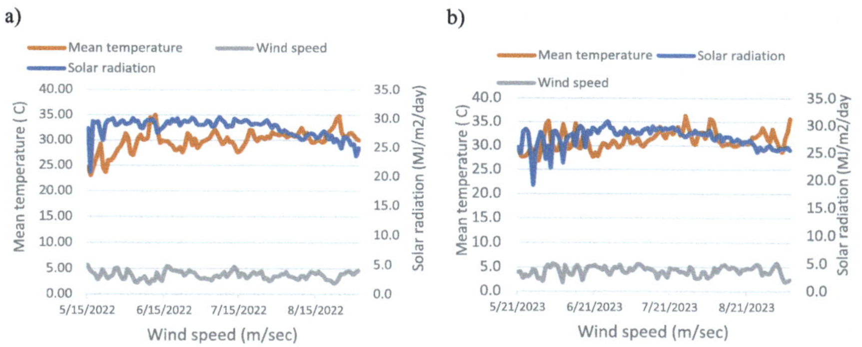

Fig. 8.2 Average temperature, solar radiation, and wind speed during the first growing season (**a**) and the second growing season (**b**)

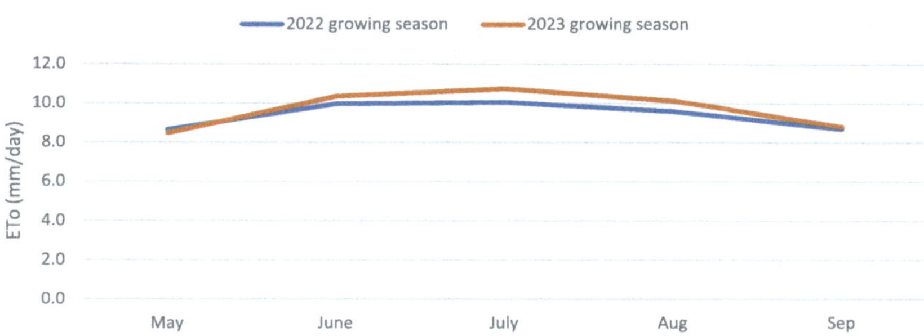

Fig. 8.3 Monthly ETo values in both growing seasons

Calculation of Applied Irrigation Water and Water Consumption by Cowpea–Maize Intercropping Systems

Utilized Irrigation Water

The calculated irrigation water volumes for crop evapotranspiration or irrigation under optimal conditions employing a drip system were determined using the subsequent equation (Allen et al., 1998):

$$\mathrm{ETc} = \mathrm{ETo} \times \mathrm{kc} \tag{8.1}$$

where ETc represents crop evapotranspiration (mm/day), ETo denotes reference evapotranspiration (mm/day), and kc signifies the crop coefficient.

ETo values were calculated using the Penman–Monteith equation (Allen et al., 1998). The kc values for maize, as documented by Taha and Kasem (2022), are 0.35, 0.98, 1.12, and 0.45 for the initial, developmental, middle, and end seasons, respectively, and were employed to ascertain the requisite irrigation water depth in accordance with Allen et al. (1998).

The requisite irrigation water volume was determined using the subsequent equation (Vermeiren & Jopling, 1984):

$$\mathrm{AIW} = \frac{\mathrm{ETo} \times I}{\mathrm{Ea}\left(1 - \mathrm{LR}\right)} \tag{8.2}$$

AIW represents the applied irrigation water depth (mm/day), whereas Ea denotes the application efficiency, which is 85% for the drip irrigation method. *I* is the time interval in days and LR is the leaching requirements, which is not considered due to low soil salinity level. We presumed that the irrigation water supplied to the intercropped maize suffices to meet the water requirements of its intercropping systems with cowpea, as reported by Taha and Kasem (2022). Irrigation was performed at scheduled intervals of every 2 days via the drip lateral lines. Four rows on the upper section of the raised beds for intercropping systems and sole cowpea, and two rows for sole maize. These lines are linked to the sub-main line. Each lateral line measures 20 m in length and is positioned 0.70 m apart on the sub-main, with integrated emitters with a discharge rate of 2 L/h, spaced 0.30 m along the lateral lines. A differential pressure tank was linked to the drip irrigation system to provide fertilizer through the irrigation water.

Water Consumptive Use

Measurements of water consumptive use (WCU, mm) were acquired utilizing a time domain reflectometry sensor (TDR). Volumetric soil moisture levels at a depth of 60 cm were assessed before and after each irrigation session. The seasonal WCU values were calculated using the formula proposed by Israelsen and Hansen (1962) as follows:

$$\mathrm{WCU} = \sum_{i=1}^{i=3} \frac{\theta 2 - \theta 1}{100} \times d \tag{8.3}$$

where i = number of soil layers, $\theta 2$ = soil moisture content after irrigation (%, by volume), $\theta 1$ = soil moisture content just before irrigation (%, by volume), and d = depth of the soil layer (mm).

The Regulatory Service of the Cowpea–Maize Intercropping System in Microclimate Regulation

It has been demonstrated that intercropping with legumes can improve climate regulation by modifying the microclimate inside the field. This can have an effect on a variety of elements, including light intensity, temperature (Akchaya et al., 2025), humidity, desiccating winds, and other climatic variables (Gebru, 2015), which consequently reflected on the required irrigation amounts and the consumed water. Furthermore, Ghanbari et al. (2010) indicated that intercropping cowpea with maize significantly alters the canopy microclimate by affecting light availability, temperature, and soil moisture levels. The benefits of cowpea–maize intercropping, according to Dube et al. (2014), could be due to increased moisture retention compared to monoculture farming providing, which represents an advantage during periods of water stress. Consequently, the assessment of irrigation water applied to the examined cowpea–maize intercropping systems, along with the volume of water consumed, water equivalent ratio, and variations in water usage, serves as valuable metrics for explaining the contribution of cowpea in alleviating climate fluctuation.

Amounts of Applied Irrigation Water for Sole and Intercropped Crops

Table 8.3 delineates the anticipated volumes of irrigation water utilized in the examined treatments for both sole maize and its intercropping systems with cowpea, in addition to sole cowpea. The irrigation quantities assigned to sole maize were identical to those allotted for the intercropping system with PD2 and PD3. The designated irrigation volume for the intercropping system with PD1 included the irrigation water amount applied to cowpea 2 weeks prior to maize cultivation, together with the water allotted to maize. Table 8.3 indicates that the irrigation water utilized for both sole and intercropped crops was higher in the second growing season compared to the first, due to increased seasonal ETo values (Fig. 8.3).

Water stress treatments of 100% and 80% ETc led to a 19% and 40% reduction in irrigation water applied to maize and its intercropping systems during the first growing season, respectively, in comparison to the required irrigation quantities. During the second growing season, the reductions were 17% and 34%, respectively, in relation to the required irrigation quantities. Irrigation water applied to sole maize and sole cowpea diminished by 19% and 39%, respectively, while employing 100% and 80% ETc during the first growing season, in comparison to the required irrigation levels. During the second growing season, the amounts of water applied were reduced by 16% and 36% with the application of 100% and 80% ETc, respectively, compared to the required irrigation volumes (Table 8.3). Taha and Kasem (2022) reported that the water applied to the cowpea–maize intercropping system under surface irrigation was reduced by 15% and 28%, corresponding to 100% and 80% ETo, respectively, in Giza Governorate, Egypt, averaged over two growing seasons.

Water Consumption by Sole and Intercropped Crops

Table 8.4 indicated that the water consumption values for maize intercropping systems exceeded those of sole maize owing to the inclusion of intercropped cowpea. This result demonstrated that water loss by evapotranspiration and/or deep percolation decreased with the increase of ground cover from intercropped cowpea. The peak water consumption occurred when cowpea was sown 2 weeks prior to maize, due to the substantial irrigation applied, which created enhanced ground cover before maize planting, thereby minimizing evaporation losses, as noted by Zhang et al. (2022). Reports suggest that while intercropping may lead to increased total water consumption compared to sole cropping, it demonstrates greater water use efficiency, especially when considering the improved yield (Wang et al., 2015). The improved utilization of the applied water presented in Table 8.4 is also resulting in reduced water losses due to deep percolation within this intercropping system. The extensive ground cover provided by cowpea is also credited with reducing soil

Table 8.3 Utilized irrigation water (mm) for sole maize, sole cowpea, and their intercropping systems throughout various cowpea planting dates

	First growing season			Second growing season		
	RI	WS_1	WS_2	RI	WS_1	WS_2
Sole maize	1193	961	710	1271	1026	771
Sole cowpea	665	501	362	708	563	381
Intercropped cowpea in PD1	1264	1020	757	1340	1081	811
Intercropped cowpea in PD2	1193	961	710	1271	1026	771
Intercropped cowpea in PD3	1193	961	710	1271	1026	771

RI, WS_1, and WS_2 = irrigation with 120, 100, and 80% ETc, respectively; PD1 = cowpea sown 2 weeks prior to maize cultivation; PD2 = cowpea sown concurrently with maize cultivation; PD3 = cowpea sown 2 weeks subsequent to maize cultivation

Table 8.4 Water consumptive use (mm) of sole maize, sole cowpea, and their intercropping systems throughout various cowpea planting dates

	First season			Second season		
	RI	WS_1	WS_2	RI	WS_1	WS_2
Sole maize	916	799	625	1131	893	663
Sole cowpea	599	446	319	630	490	328
Intercropped cowpea in PD1	1038	890	649	1093	938	689
Intercropped cowpea in PD2	998	875	639	1011	913	671
Intercropped cowpea in PD3	998	865	639	1011	913	671

RI, WS_1, and WS_2 = irrigation with 120, 100, and 80% ETc, respectively; PD1 = cowpea sown 2 weeks prior to maize cultivation; PD2 = cowpea sown concurrently with maize cultivation; PD3 cowpea sown 2 weeks subsequent to maize cultivation

evaporation. Furthermore, the contrasting rooting patterns of cowpea and maize (deep versus shallow roots) enhanced access to soil moisture at various depths and allowed for the more extensive utilization of soil volume, hence optimizing water usage efficiency, as noted by Gebru (2015). He et al. (2011) indicated that intercropping could result in diminished daytime relative humidity and heightened nocturnal relative humidity, hence facilitating less water loss from plants during the day and the maintenance of moisture levels at night.

A study by Ghanbari et al. (2010) revealed that intercropping cowpea with maize enhances soil moisture conservation, reduces water evaporation, and increases light interception compared to maize monoculture, indicating the influence of intercropped cowpea on the maize microclimate. Furthermore, due to their extensive leaf area and early canopy formation, cowpeas have demonstrated a considerable capacity for water conservation, hence providing enhanced potential to mitigate the adverse impacts of moisture stress (Layek et al., 2018). Dhakal et al. (2016) assert that intercropping cowpea with maize enhances water availability by augmenting soil organic matter, hence mitigating soil degradation.

Moreover, increased water consumption was seen in the second growing season compared to the first, due to rising monthly ETo values (Fig. 8.3).

The Productivity of Sole and Intercropped Maize Crops across Different Irrigation Regimes

Sole and Intercropped Maize

Significant differences ($p < 0.05$) were noted between the yield of sole maize and its yield in the three intercropping systems, as well as across the three irrigation regimens in the first growing seasons. Whereas, in the second growing season, there were significant differences ($p < 0.05$) between the yield of sole maize and its yield in the three intercropping systems, as well as across the three irrigation regimens and the interaction between irrigation treatments and cropping systems (Table 8.5).

Table 8.5 indicates that intercropped maize production increased in the second growing season relative to the first, irrespective of irrigation volume or cowpea planting date. This enhancement is ascribed to the residual effects of cowpea as a leguminous crop on soil health, resulting from biological nitrogen fixation by the legume in the second season compared to the first (Ouda et al., 2020). Shen and Chu (2004) asserted that legumes can transfer fixed nitrogen, a crucial ingredient for intercropped cereals during their concurrent growth phase. Moreover, Collins et al. (2017) indicated that, within an intercropping system, the overall plant population can occasionally surpass that of the base crop in a monoculture, thereby enabling a greater number of plants to utilize the available resources. Thus, it permits more biomass to be created due to more assimilated production by the intercrops together.

Moreover, the findings in Table 8.3 indicated a decline in sole maize yield during the second season compared to the first, attributed to the depleting effect of maize as a cereal crop on soil nutrients. This observation aligns with Fujisao et al. (2020), who noted that a reduction in maize yield correlates with soil degradation resulting from continuous cultivation. The yield of intercropped maize was inferior to that of sole cultivation when cowpea was sown 2 weeks prior to maize in both growing seasons and across all irrigation treatments. This may be due to interspecific competition for light and soil nutrients between cowpea and maize, with cowpea rapidly establishing ground cover earlier by 2 weeks before maize cultivation, as shown by Megawer et al. (2010). The yield of intercropped maize was higher than that of sole maize cultivation when cowpea was sown on the same day as maize planting and when cowpea was intercropped 2 weeks subsequent to maize cultivation. This

Table 8.5 The maize yields (ton/ha) in sole and intercropped systems with varying cowpea planting dates (CS) and irrigation treatments (IR) during both growing seasons

Cropping system	First growing season				Second growing season			
	RI	WS_1	WS_2	Mean	RI	WS_1	WS_2	Mean
Sole maize	6.90	5.93	4.15	5.66	6.64	5.63	3.85	5.37
Intercropped cowpea in PD1	5.56	4.75	3.29	4.53	5.68	4.81	3.22	4.57
Intercropped cowpea in PD2	7.21	6.06	4.18	5.82	7.42	6.19	4.12	5.91
Intercropped cowpea in PD3	7.34	6.23	4.25	5.94	7.76	6.52	4.35	6.21
Mean	6.75	5.74	3.97		6.88	5.79	3.89	
$LSD_{0.05}$								
IR	1.36				0.63			
CS	1.03				1.34			
IR X CS	NS				2.33			

RI, WS_1, and WS_2 = irrigation with 120, 100, and 80% ETc, respectively; PD1 = cowpea sown 2 weeks prior to maize cultivation; PD2 = cowpea sown concurrently with maize cultivation; PD3 = cowpea sown 2 weeks subsequent to maize cultivation; NS = non-significant

outcome is ascribed to minimal interspecific competition for light and soil nutrients between cowpea and maize, enabling maize plants to utilize available resources well. A consistent pattern was noted throughout the three irrigation treatments in both growing seasons (Table 8.5). Ouda et al. (2020) and Zohry et al. (2017) achieved analogous results. Mishra (2019) documented an enhancement in both cowpea and maize inside its intercropping system, in contrast to the sole production of each crop.

The production of sole maize decreased under conditions of water scarcity, with irrigation with 100% and 80% ETc resulting in losses of 14% and 40% in yield, respectively, during the first growing season. During the second growing season, the decrease was 15% and 42% under irrigation with 100% and 80% ETc, respectively. A more significant reduction in the yield of intercropped maize was observed when cowpea was sown simultaneously with maize or 2 weeks later, both subjected to water stress treatments, due to interspecific competition for water between the cowpea and maize plants. Nevertheless, intercropping cowpea 2 weeks earlier diminishes maize yield losses due to reduced interspecific competition for water. These results were consistent over both growing seasons (Table 8.5).

Sole and Intercropped Cowpea

Table 8.6 indicates significant differences *($p < 0.05$)* in cowpea yields between sole and intercropped systems across the three irrigation treatments, as well as the interaction effects of irrigation treatments and cropping systems in both growing seasons.

Moreover, Table 8.6 illustrates that, with the required irrigation, the yield of sole cowpea was higher than that of intercropped cowpea in both growing seasons due to the interspecific competition between cowpea and maize plants under intercropping, with an enhancement in the second season relative to the first, ascribed to cowpea residual effect in the soil. The yield of intercropped cowpea was maximized when planted 2 weeks prior to maize cultivation and minimized when planted 2 weeks subsequent to maize cultivation in both growing seasons. Eskandari (2012) asserts that intercropping cowpea with maize significantly affects cowpea due to its low stature and limited capacity to compete for resources. Interspecific competition between maize and cowpea plants was most pronounced when cowpeas were sown 2 weeks after maize, and least pronounced when sown 2 weeks before maize, as elucidated by Abou-Keriasha et al. (2011).

Under the applied water stress treatments, the yield of sole cowpea decreased by 9% and 32% for those irrigated at 100% and 80% ETc, respectively, averaged over both growing seasons, compared to yields achieved under optimal irrigation conditions (Table 8.6). Elevated yield reductions were noted for intercropped cowpea under both water stress conditions, likely due to interspecific competition for water between cowpea and maize. Vasconcelos et al. (2020) indicated that cowpea

Table 8.6 The sole and intercropped cowpea yields (ton/ha) under its intercropping systems with maize (CS) and irrigation treatments (IR) over both growing seasons

Cropping system	First growing season				Second growing season			
	RI	WS_1	WS_2	Mean	RI	WS_1	WS_2	Mean
Sole cowpea	58.08	53.04	41.04	50.72	59.48	54.34	42.24	52.02
Intercropped cowpea in PD1	48.24	43.44	33.36	41.68	49.54	44.64	34.46	42.88
Intercropped cowpea in PD2	36.24	31.68	24.24	30.72	37.44	32.78	25.24	31.82
Intercropped cowpea in PD3	33.12	28.80	21.84	27.92	34.22	29.80	22.74	28.92
Mean	43.92	39.24	30.12	37.76	45.17	40.39	31.17	38.91
$LSD_{0.05}$								
IR	0.65				1.96			
CS	0.73				1.06			
IR X CS	1.27				1.83			

RI, WS_1, and WS_2 = irrigation with 120, 100, and 80% ETc, respectively; PD1 = cowpea sown 2 weeks prior to maize cultivation; PD2 = cowpea sown concurrently with maize cultivation; PD3 = cowpea sown 2 weeks subsequent to maize cultivation

possesses moderate drought tolerance; however, our findings revealed a significant decrease in its intercropped yield, particularly when irrigated at 80% ETc. Mugi-Ngenga et al. (2023) noted that the complementary utilization of irrigation water arises from the differing root systems, specifically fibrous roots in cereals and taproots in legumes. Nonetheless, this complementarity may transition to competition under conditions of water stress (Senghor et al., 2023).

Evaluation Metrics for Assessing the Role of Cowpea in Climate Regulation

Land Equivalent Ratio

To evaluate the efficiency of the studied cowpea–maize intercropping systems in land use, land equivalent ratio (LER) was utilized. LER is an assessment of the land use efficiency of the intercropping system (Rao & Willey, 1980) and is computed as follows:

$$LER = LER_A + LER_B = \frac{YintA}{YmonoA} + \frac{yintB}{YmonoB} \tag{8.4}$$

where LER_A and LER_B are the partial land equivalent ratios of crop A and crop B, respectively. $Y_{int,A}$ and $Y_{int,B}$ are the intercropped yields of crop A and crop B, respectively. $Y_{mono,A}$ and $Y_{mono,B}$ are the monoculture yields of crop A and crop B, respectively. If $LER > 1$, it indicates a higher land utilization efficiency of the intercropping system than that of the monoculture.

Table 8.7 displays the values of partial LER for cowpea (LER_{cowpea}), partial LER for maize (LER_{maize}), and total LER (LER_{total}) across both growth seasons. The table results demonstrated that the values of LER_{cowpea} were maximized when cowpea was planted 2 weeks prior to maize, and diminished when cowpea was sown concurrently with maize or 2 weeks subsequent to maize planting. Conversely, the highest values of LER_{maize} were attained when cowpea was sown 2 weeks after maize, and these values diminished when cowpea was planted concurrently with maize or 2 weeks before maize planting. The results remain consistent throughout the three irrigation regimes in both growing seasons.

The highest LER_{total} values were recorded when cowpea was sown concurrently with maize, observed across all three irrigation regimes in both growing seasons.

Moreover, the findings indicated elevated values of LER_{cowpea}, LER_{maize}, and LER_{total} in the second growing season, attributable to increased yields of both crops during this period, in contrast to the first growing season, when irrigated adequately. Abou-Keriasha et al. (2011) intercropped cowpea with maize at 100% of the recommended planting density for both crops across three planting dates: prior to, simultaneous with, and subsequent to maize cultivation. They observed comparable outcomes, with the highest LER_{cowpea} recorded for early cowpea planting, while the peak values of LER_{maize} and LER_{total} were achieved with late cowpea planting. Moreover, these values were elevated in the second growing season in comparison to the first growth season.

Similar total LER values were documented in Egypt by El-Ghobashy et al. (2018), Elsaid et al. (2019), El-Shamy et al. (2022), El-Ghobashy et al. (2020), Salama et al. (2022), and Shehata and Hamd-Alla (2023), across various cropping patterns of cowpea and maize, as well as fertilizer treatments.

The implementation of the two water stress treatments led to a decrease in the values of LER_{cowpea}, LER_{maize}, and LER_{total} due to diminished yields of both

Table 8.7 The land equivalent ratios (LER) influenced by maize intercropping systems (IS) under the examined irrigation treatments (IR) across both growing seasons

Irrigation treatment	Intercropping systems	First growing season			Second growing season		
		LER_{cowpea}	LER_{maize}	LER_{total}	LER_{cowpea}	LER_{maize}	LER_{total}
RI	PD1	0.83	0.81	1.64	0.83	0.86	1.69
	PD2	0.62	1.04	1.67	0.63	1.12	1.75
	PD3	0.57	1.06	1.63	0.58	1.17	1.74
WS1	PD1	0.81	0.80	1.61	0.82	0.85	1.68
	PD2	0.60	1.02	1.62	0.60	1.11	1.71
	PD3	0.54	1.05	1.59	0.54	1.16	1.70
WS2	PD1	0.81	0.79	1.61	0.82	0.84	1.65
	PD2	0.59	1.03	1.62	0.60	1.07	1.67
	PD3	0.53	1.02	1.56	0.54	1.12	1.66

RI, WS_1, and WS_2 = irrigation with 120, 100, and 80% ETc, respectively; PD1 = cowpea sown 2 weeks prior to maize cultivation; PD2 = cowpea sown concurrently with maize cultivation; PD3 = cowpea sown 2 weeks subsequent to maize cultivation

intercrops across both growing seasons. Nonetheless, the implementation of the two water stress treatments during the second growing season yielded greater LER_{total} values compared to those recorded under the three irrigation treatments in the first growing season. These findings demonstrate the contribution of cowpea in enhancing land and resource use efficiency, boosting soil fertility through its nitrogen-fixing capability, reducing soil evaporation through its ground cover and the potential for increased total yields and economic returns relative to monocropping.

Water Equivalent Ratio

Water equivalent ratio (WER) quantifies the water volume necessary for individual crops to achieve yields comparable to those obtained with one unit of water in intercrops, as determined by the formula proposed by Mao et al. (2012):

$$\begin{aligned} \text{WER} = & \left[\left(\text{Yint,A} / \text{WUint} \right) / \left(\text{Ymono,A} / \text{WUmono,A} \right) \right] \\ & + \left[\left(\text{Yint,B} / \text{WUint} \right) / \left(\text{Ymono,B} / \text{WUmono,B} \right) \right] \end{aligned} \tag{8.5}$$

where $Y_{int,A}$, $Y_{mono,A}$, $Y_{int,B}$, and $Y_{mono,B}$ are the yields of intercropped crop A, monocrop A, intercropped crop B (legume crop), and monocrop B, respectively. WU_{int}, $WU_{mono,A}$, and $WU_{mono,B}$ are water use by the intercropping system, water use of monocrop A, and water use of intercropped crop B, respectively. If the WER > 1, it suggests that the production of a unit of water is higher than that of the monoculture.

Change in Water Use

The change in water use by the intercrops (ΔWU) index developed by Morris and Garrity (1993) was employed alongside WER to evaluate the benefits of water use efficiency in intercrops relative to monoculture. ΔWU measures the relative difference between the observed water intake in the intercrop (WU_{int}, obs) and the anticipated water use derived from the water consumption of the two crop species, adjusted by weights reflecting their respective proportions in the intercrop (WU_{int}, exp). The anticipated water consumption in intercropping correlates with the water utilization of each species in monoculture and the relative yield achieved in intercropping compared to monoculture (partial LER). Consequently, this formula employs the partial land equivalent ratio of crops A and B as coefficients for ΔWU (Mao et al., 2012), as delineated below:

$$\Delta WU = \left[\text{WUint,obs} / \text{WUint,exp} \right] - 1 \tag{8.6}$$

$$\Delta WU = \left[\text{WUint} / \left(\text{LERint,A x WUmono,A}\right) + \left(\text{LERint,B x WUmono,B}\right)\right] - 1 \quad (8.7)$$

Both WER and ΔWU evaluate whether a particular intercrop yield (consisting of two species) would be achieved with augmented water (WER > 1; ΔWU < 0) or diminished water (WER < 1; ΔWU > 0) relative to that employed in monocultures.

The results for WER_{cowpea}, WER_{maize}, and WER_{total} are displayed in Table 8.8. The peak WER_{cowpea} values across the three irrigation treatments were recorded when cowpea was sown prior to maize in both growing seasons. The peak values of WER_{maize} and WER_{total} were recorded when cowpea was cultivated following maize during both growth seasons. Moreover, the values of partial and total WER were heightened during the second growing season relative to those in the first growing season. The increased WER_{total} in the second growing season, attributed to the residual effects of cowpea on the soil, improved soil water retention, hence positively affecting the yield of intercropped maize. Intercropping has been shown to markedly enhance the water use efficiency of the system, indicating a considerable advantage in water utilization (Mao et al., 2012). Taha and Kasem (2022) intercropped 50% of cowpea recommended planting density with 100% maize recommended planting density, achieving a WER value of 1.26.

Moreover, the application of water stress treatments resulted in diminished partial and total WER values during both growing seasons. Nevertheless, the WER_{total} values under WS1 during the second growing season exceeded those under RI in the first growing season (Table 8.8). Consequently, intercropping with legumes can improve water usage efficiency for maize, particularly under conditions of moderate water stress as stated by Liu et al. (2025).

Table 8.8 demonstrates that the three intercropping systems attained total WER values above 1 and ΔWU values below 0, even under water stress conditions,

Table 8.8 The water equivalent ratios (WER) and change in water use (ΔWU) influenced by maize intercropping systems (IS) under the studied irrigation treatments (IR) in both growing seasons

		First growing season				Second growing season			
IR	IS	WER_{cowpea}	WER_{maize}	WER_{total}	ΔWU	WER_{cowpea}	WER_{maize}	WER_{total}	ΔWU
RI	PD1	0.48	0.71	1.19	−0.17	0.48	0.89	1.37	−0.26
	PD2	0.37	0.96	1.33	−0.17	0.39	1.25	1.64	−0.29
	PD3	0.34	0.98	1.32	−0.14	0.36	1.31	1.67	−0.27
WS1	PD1	0.41	0.70	1.11	−0.11	0.43	0.81	1.24	−0.19
	PD2	0.30	0.91	1.21	−0.05	0.32	1.05	1.37	−0.15
	PD3	0.28	0.95	1.23	−0.03	0.29	1.10	1.40	−0.14
WS2	PD1	0.40	0.79	1.19	−0.15	0.39	0.80	1.19	−0.15
	PD2	0.29	1.01	1.30	−0.07	0.29	1.03	1.32	−0.10
	PD3	0.27	1.02	1.29	−0.03	0.26	1.09	1.35	−0.08

RI, WS_1, and WS_2 = irrigation with 120, 100, and 80% ETc, respectively; PD1 = cowpea sown 2 weeks prior to maize cultivation; PD2 = cowpea sown concurrently with maize cultivation; PD3 = cowpea sown 2 weeks subsequent to maize cultivation.

signifying improved water use efficiency. Ouda and Zohry (2023) observed similar WER and ΔWU values in Egypt for sugar beet intercropping systems with legumes, whereas Zohry and Ouda (2025) documented analogous values for wheat intercropping systems with leguminous crops.

Reports suggest that maize possesses a superficial root system, whereas cowpea produces a deeper root structure, enabling both crops to extract water from varying depths within the soil profile. The root system plays a crucial role in water acquisition in intercrops, as species mixtures have been shown to increase root length density, hence decreasing water loss through evaporation (Pankou et al., 2021). This dynamic diminishes direct competition for identical water resources and enhances the overall water use efficiency of the system (Dube et al., 2014). Evidence indicates that intercropping cowpea with maize effectively conserves soil moisture content, potentially reaching 16% during the active crop development phase and 17% post-harvest (Ayele, 2020). Intercropping has been shown to enhance water use efficiency by 4% to 99% compared to sole crops, especially under conditions of unconstrained water availability (Mao et al., 2012).

Evaluation of Economic Performance

Total income (TI) and monetary advantage index (MAI) were computed to assess the economic benefits of cowpea–maize intercropping systems in comparison to sole planting, utilizing the equation presented by Raza et al. (2021):

$$\text{Total return}\left(\text{US\$ ha}^{-1}\right) = \left(\text{yield A} \times \text{price A}\right) + \left(\text{yield B} \times \text{price B}\right) \quad (8.8)$$

where A and B denote sole and intercropped species, respectively. The prices utilized in the analysis were obtained from the farm prices as follows: The prices per ton for cowpea and maize in the first season were 184.76 and 32.13 USD/ton, respectively, while in the second season, the prices were 315.44 and 27.42 USD/ton, respectively.

Additionally, the monetary advantage index (MAI) was derived from the LER values. The analysis elucidates the economic benefits of the intercropping system, with a positive MAI value signifying its advantages. The MAI was computed following the approach defined by Ghosh (2004).

$$\text{MAI} = \left(\text{Value of combined intercrops}\right)^{*}\left(\text{LER} - 1\right)/\text{LER} \quad (8.9)$$

Table 8.9 exhibited increased total income figures for all assessed intercropping systems relative to maize monoculture, with the highest values attained when cowpea was intercropped 2 weeks prior to maize cultivation across the three irrigation regimes in the first growing season. Conversely, in the second growing season, the maximum total income values were achieved when cowpea was intercropped

2 weeks after maize cultivation across all three irrigation regimes. The result was attributed to higher maize price in the second season compared to the first season. This suggests that the combined yield and income from the intercropping systems exceed those from sole maize cultivation.

Furthermore, positive values of MAI were attained in both growing seasons. The values of MAI followed the same trend as total income, where it was the highest when cowpea was intercropped 2 weeks prior to maize cultivation across the three irrigation regimes in the first growing season. In the second growing season, the maximum MAI values were achieved when cowpea was intercropped 2 weeks after maize cultivation across all three irrigation regimes (Table 8.9).

These findings correspond with those of LER, where Dhima et al. (2007) demonstrated that the greatest LER values in an intercropping system yield economic benefits evident in MAI values. The outcomes can be ascribed to heightened overall productivity in intercropping with somewhat reduced input expenditure (Banik et al., 2006). Ouda and Zohry (2026) achieved similar findings when intercropping soybean with maize at varying soybean planting densities.

The Regulatory Service of the Cowpea–Maize Intercropping System on Soil Fertility and Nutrient Circulation

As a leguminous crop, cowpea can enhance organic matter and soil nutrients through rhizodeposition and symbiotic nitrogen fixation (Sánchez-Navarro et al., 2019). In addition to adding nitrogen to the soil, cowpeas' roots emit carbon-rich molecules that nourish soil bacteria, which helps future crops absorb and use other nutrients, such as phosphorus (Amorim et al., 2022).

Table 8.9 Total income (TI) and monetary advantage index (MAI) of the studied maize intercropping systems under different irrigation treatments in both growing seasons

	RI		WS_1		WS_2	
	TI (USD/ha)	MAI	TI (USD/ha)	MAI	TI (USD/ha)	MAI
First growing season						
Sole maize	1275	—	1096	—	767	—
Intercropped cowpea in PD1	2577	1006	2273	861	1680	636
Intercropped cowpea in PD2	2497	1002	2138	818	1551	594
Intercropped cowpea in PD3	2420	935	2077	771	1487	534
Second growing season						
Sole maize	2095	–	1776	–	1214	—
Intercropped cowpea in PD1	3150	1286	2741	1110	1961	772
Intercropped cowpea in PD2	3367	1443	2851	1184	1992	799
Intercropped cowpea in PD3	3386	1440	2874	1183	1996	793

RI, WS_1, and WS_2 = irrigation with 120, 100, and 80% ETc, respectively; PD1 = cowpea sown 2 weeks prior to maize cultivation; PD2 = cowpea sown concurrently with maize cultivation; PD3 = cowpea sown 2 weeks subsequent to maize cultivation.

Soil Organic Matter Content

In this investigation, the level of soil organic matter was assessed prior to cultivation and following the harvest of sole and intercropped maize at the conclusion of the second season by collecting soil samples from three distinct strata (0–20, 20–40, and 40–60 cm). The carbon content was quantified using wet oxidation (Walkley & Black, 1934). The organic matter content was determined by multiplying organic carbon by 1.724.

Table 8.10 displays the quantified values of soil organic matter following the harvest of sole maize and its intercropping systems throughout the three examined irrigation treatments. The quantified percentage of soil organic matter before cultivation was 0.31%. The table indicated that the quantified levels of soil organic matter were the lowest during sole maize farming. Balík et al. (2020) assert that ongoing maize farming can diminish soil organic matter due to heightened breakdown and reduced carbon input. The rapid decomposition of maize as a C4 plant residue, coupled with intensive methods such as frequent tillage, significantly accelerates the depletion of soil organic matter.

The table's results indicated that the percentage increase in soil organic matter correlates with the duration of the cowpea growing season, with the most significant increase occurring when cowpea is planted 2 weeks prior to maize cultivation, and the least increase observed when cowpea is planted 2 weeks subsequent to maize cultivation. This tendency was noted with the application of required irrigation quantities. Whereas, under water stress treatments, a similar trend was seen with a reduction in the values of soil organic matter growth (Table 8.10). Research indicated that exposing cowpea to water stress can indirectly contribute to a reduction in soil organic matter over time by negatively impacting the plant's growth and the amount of biomass it produces and returns to the soil (Drăghici et al., 2024).

Table 8.10 Soil organic matter (SOM%) following the harvest of sole maize and its intercropping systems with cowpea, along with the percentage variation in soil organic matter under the examined irrigation regimens

	SOM after harvest			Percentage variation in SOM		
	RI	WS_1	WS_2	RI	WS_1	WS_2
Sole maize	0.29	0.27	0.23	−6	−13	−26
Intercropped cowpea in PD1	0.38	0.36	0.34	23	16	10
Intercropped cowpea in PD2	0.37	0.35	0.33	19	13	6
Intercropped cowpea in PD3	0.36	0.34	0.32	16	10	3

RI, WS_1, and WS_2 = irrigation with 120, 100, and 80% ETc, respectively; PD1 = cowpea sown 2 weeks prior to maize cultivation; PD2 = cowpea sown concurrently with maize cultivation; PD3 = cowpea sown 2 weeks subsequent to maize cultivation

Nutrient Circulation

Soil samples were obtained prior to planting and post-harvest to ascertain total nitrogen concentration. Twelve cores were randomly extracted from each treatment with a hand-operated auger, and the procedure was reiterated. Soil samples were air-dried, meticulously crushed, and subsequently sieved through a 2 mm mesh to get homogeneous representative samples. Nitrogen was quantified using the micro-Kjeldahl method (Hesse, 1971). The quantification of available phosphorus in soil was conducted using stannous chloride (SnCl2) using UV-Vis spectrophotometry (Jackson, 1973).

Soil Available Nitrogen

Table 8.11 presents the impact of sole maize, its intercropping systems, and irrigation interventions on soil accessible nitrogen. The findings indicated significant differences *($p < 0.05$)* in nitrogen levels between sole and intercropped systems across the three irrigation treatments, as well as the interaction effects of irrigation treatments and cropping systems in both growing seasons.

Table 8.11 indicated that the concentrations of accessible nitrogen in the soil were higher in all maize intercropping systems compared to sole maize irrigated with the three irrigation regimes. The highest concentrations of soil nitrogen were seen when cowpea was sown 2 weeks prior to maize. Given maize's substantial nitrogen demands, intercropping with legumes can enhance nitrogen uptake by maize, as cowpea fix nitrogen in the soil, making it available for maize (Nasar et al., 2021). The magnitude of biological nitrogen fixation and its corresponding contribution varies among cropping systems, soil properties, and soil management practices (Stagnari et al., 2017). It has been reported that cowpea biologically fix between 36 and 75 kg of nitrogen per hectare (Vasconcelos et al., 2020). Boukar et al. (2013) demonstrated that intercropping cowpea with cereals can enhance soil fertility by compensating for the nitrogen depletion caused by the cereals. Maize root exudates have been shown to enhance the expression of nodulation-related genes, and intercropping maize with legumes has been associated with increased nitrogen fixation, leading to improved nitrogen use efficiency in intercropping (Hu et al., 2021).

Table 8.11 reveals that increased levels of soil accessible nitrogen were noted in the second growing season, in contrast to those documented in the first growing season, due to the residual effect of cowpea as a leguminous crop. Ghosh et al. (2007) asserted that cowpea intercropped with maize positively influenced the productivity of subsequent crops. Moreover, Ouda et al. (2020) documented an 88% increase in soil nitrogen following the harvest of intercropped maize with cowpea after a two-year study.

Table 8.11 Soil nitrogen availability (mg/kg) influenced by sole maize, intercropping systems (CS), and irrigation treatments (IR) over both growing seasons

Cropping systems	First growing season				Second growing season			
	RI	WS_1	WS_2	Mean	RI	WS_1	WS_2	Mean
Sole maize	11.39	10.00	8.96	10.12	11.99	10.12	9.02	10.38
Intercropped cowpea in PD1	14.39	12.79	10.87	12.68	14.49	13.04	11.33	12.95
Intercropped cowpea in PD2	13.30	11.82	10.23	11.78	13.80	12.04	10.73	12.19
Intercropped cowpea in PD3	12.20	10.28	9.35	10.61	12.40	10.54	9.74	10.89
Mean	12.82	11.22	9.85		13.17	11.44	10.21	
$LSD_{0.05}$								
IR	0.27				0.27			
CS	0.42				0.39			
IR X CS	0.74				0.68			

RI, WS_1, and WS_2 = irrigation with 120, 100, and 80% ETc; respectively; PD1 = cowpea sown 2 weeks prior to maize cultivation; PD2 = cowpea sown concurrently with maize cultivation; PD3 = cowpea sown 2 weeks subsequent to maize cultivation

Furthermore, the levels of accessible nitrogen in the soil decreased under both water stress regimes (Table 8.11). Melo et al. (2025) indicated that reduced soil water availability diminished both the quantity and mass of cowpea nodules, thereby impairing biological nitrogen fixation due to decreased accumulation of allantoin, allantoic acid, amino acids, nitrate, and ammonia in the plant's shoots and roots. In both water stress regimes, the available soil nitrogen levels under maize intercropping systems surpassed those observed under sole maize farming during both growing seasons (Table 8.11), hence underscoring the benefits of intercropping leguminous crops such as cowpea with maize.

Soil Available Phosphorus

Table 8.12 illustrates the effect of sole maize, its intercropping systems, and irrigation practices on soil available phosphorus. The results in the table indicated significant differences *($p < 0.05$)* in soil accessible phosphorus among the intercropping systems and irrigation treatments across both seasons, as well as their interaction.

Similar results to the available soil nitrogen were noted for available soil phosphorus, which exhibited increased levels across all maize intercropping systems compared to sole maize irrigated under the three irrigation regimes. The highest concentrations of soil available phosphorus were seen when cowpea was grown 2 weeks prior to maize (Table 8.12). This enhancement is frequently attributed to cowpea roots exuding organic acids that increase phosphorus solubility (Dahmardeh et al., 2010). The augmentation of phosphorus availability correlated with notable pH alterations in the cowpea rhizosphere within intercropping systems, exhibiting greater rhizosphere acidification than in sole cropping, alongside heightened soil respiration in intercropping relative to sole cropping (Latati et al., 2014). Gustiningsih et al. (2023) asserts that cowpea may flourish in acidic soils, even in

Table 8.12 Available soil phosphorus (mg/kg) influenced by sole maize, intercropping systems (CS), and irrigation treatments (IR) over both growing seasons

Cropping systems	First growing season				Second growing season			
	RI	WS_1	WS_2	Mean	RI	WS_1	WS_2	Mean
Sole maize	4.11	3.89	3.65	3.88	3.21	2.98	2.75	2.98
Intercropped cowpea in PD1	5.88	5.19	4.17	5.08	6.38	5.69	4.57	5.55
Intercropped cowpea in PD2	5.52	4.73	3.91	4.72	6.01	5.22	4.32	5.18
Intercropped cowpea in PD3	5.17	4.29	3.81	4.42	5.68	4.78	4.21	4.89
Mean	5.17	4.53	3.89		5.32	4.67	3.96	
$LSD_{0.05}$								
IR	0.07				0.37			
CS	0.09				0.34			
IR X CS	0.16				0.58			

RI, WS_1, and WS_2 = irrigation with 120, 100, and 80% ETc, respectively; PD1 = cowpea sown 2 weeks prior to maize cultivation; PD2 = cowpea sown concurrently with maize cultivation; PD3 = cowpea sown 2 weeks subsequent to maize cultivation

phosphorus-deficient situations, owing to its external adaptation mechanism including the production of organic acids (malic acid and oxalic acid) from its roots to endure low-phosphorus acidic environments. These organic acids promote the solubilization of phosphorus bound to soil particles, hence increasing its availability for plant uptake (Wang et al., 2025).

Furthermore, increased levels of soil accessible phosphorus were noted in the second growing season, unlike those documented in the first growing season, due to the residual effect of cowpea (Table 8.12). Cowpea can beneficially affect subsequent crops by liberating phosphorus from their residues, which may lead to increased yields in the following season (Gaspar et al., 2018).

Furthermore, the quantities of soil accessible phosphorus decreased under both water stress treatments (Table 8.12). Water stress adversely affects soil phosphorus release by diminishing the cowpea's capacity to absorb it, mostly due to decreased water availability, essential for the dissolution and transport of phosphorus in the soil (Moursi et al., 2013).

Nevertheless, intercropping, in conjunction with both water stress treatments, led to increased soil accessible phosphorus levels compared to those recorded under sole maize in both growing seasons (Table 8.12), hence augmenting the benefits of intercropping leguminous crops like cowpea with maize.

Cowpea–Maize Intercropping System Minimizes Associated Weeds

The fresh weight of annual weeds (broadleaf and grasses) was measured during each cowpea harvest, along with the weight of collected weed specimens during agricultural procedures; all fresh weights were aggregated for each plot.

Table 8.13 indicates that the greatest biomass of annual weeds occurred in the sole cultivation of maize during both growing seasons and under all three irrigation regimens. Moreover, the biomass of annual weeds was diminished in the second season for the cowpea–maize intercropping systems, in contrast to its biomass in the first season, attributable to the enhanced growth and yield of both cowpea and maize plants. The minimal weight of annual weeds occurred when cowpea was intercropped with maize on the same planting date during both growing seasons. Postponing the cowpea intercropping date by 2 weeks after maize planting enhances the biomass of annual weeds, in comparison to the other two intercropping methods across both growth seasons.

Abou-Keriasha et al. (2011) observed similar results, indicating that intercropping cowpea with maize on the same date led to the minimal fresh weight of annual weeds. Intercropping cowpea with maize has been observed to diminish weed growth by competing for resources such as light, water, and nutrients (Eskandari & Kazemi, 2011). Research indicates substantial decreases in weed biomass and a reduction in yield loss attributed to weeds, establishing it as an effective weed management method (Katsaruware & Manyanhaire, 2009; Jamshidi et al., 2013; Suhi et al., 2022).

Cowpea Intercropping System with Maize Can Mitigate Climate Change

The cowpea–maize intercropping system is considered a valuable climate-smart agricultural approach that can both mitigate and help farmers adapt to climate change, by increasing the plants and soil resilience to climate change. In this analysis, SALTMED model (Ragab, 2002, 2015) was used to project the effect of cowpea

Table 8.13 Fresh weight of annual weeds (ton/ha) observed in sole maize, and its intercropping systems with cowpea throughout various cowpea planting dates

Cropping systems	First growing season				second growing season			
	RI	WS_1	WS_2	Mean	RI	WS_1	WS_2	Mean
Sole maize	4.04	4.35	3.38	3.92	4.34	4.55	3.67	4.18
Intercropped cowpea in PD1	0.49	0.39	0.3	0.39	0.47	0.36	0.28	0.37
Intercropped cowpea in PD2	0.21	0.18	0.22	0.20	0.18	0.16	0.18	0.17
Intercropped cowpea in PD3	0.86	1.08	0.78	0.91	0.79	0.96	0.75	0.83
Mean	1.40	1.50	1.19		1.44	1.51	1.22	
$LSD_{0.05}$								
IR	0.13				0.12			
CS	0.12				0.10			
IR X CS	0.21				0.18			

RI, WS_1, and WS_2 = irrigation with 120, 100, and 80% ETc, respectively; PD1 = cowpea sown 2 weeks prior to maize cultivation; PD2 = cowpea sown concurrently with maize cultivation; PD3 = cowpea sown 2 weeks subsequent to maize cultivation

as a companion crop sown in different planted dates on maize yield under future climate scenarios.

Calibration and Validation of SALTMED Model

The field experiment data from both growing seasons were utilized as input for the SALTMED model. The SALTMED model is a significant process-oriented crop simulation model, which employs integrated crop-soil-water processes and has demonstrated efficacy as a tool for water resources management in agriculture (Ragab, 2015). The model has been extensively utilized for various crops irrigated with diverse irrigation and fertigation techniques throughout different regions globally. A comprehensive elucidation of the SALTMED model has been presented by Ragab (2002, 2015) and Ragab et al. (2005a, b).

The data from the first growing season was utilized to calibrate the SALTMED model, whereas the data from the subsequent growing season was employed to validate the model. The calibration involved the meticulous adjustment of the "photosynthesis efficiency factor" to achieve plausible values for both sole and intercropped maize yields, while the validation was conducted by utilizing the calibrated values within the model to predict the yields of sole and intercropped maize.

The SALTMED model necessitates the input of a soil file, crop file, climate file, irrigation file, and nitrogen file. The soil data gathered over the growing seasons was utilized to compile the soil file. The model's crop file was generated utilizing the collected field data, including planting date, leaf area index, plant height, and growth stage duration. Additional input data was acquired from the literature, namely root depth, Kc, and Kcb (Allen et al. (1998)). Additionally, we employed the "Fixed Length Growth Stages" function, wherein the harvest date is established by the user. A weather file was created for each growing season, commencing from the sowing date and concluding with the harvest date.

The weather file required the input of sunshine hours (SSH); therefore, it was computed using the value of mean temperature (T_{mean}) according to Abd El-Wahed and Snyder (2015) as follows:

$$SSH = 4.352 + 0.232\left(T_{mean}\right) \tag{8.10}$$

The SALTMED model computes reference evapotranspiration (ETo) values using various approaches, including the FAO modified Penman–Monteith equation as per Allen et al. (1998). The irrigation schedule employed in the field experiment was utilized to create the irrigation files for each irrigation treatment in every growing season, incorporating fertigation amounts. The nitrogen doses utilized in each growing season were documented in the nitrogen file.

SALTMED model offers comprehensive physiological parameters embedded in the "Crop growth tab" for biomass calculation, which can be utilized or adjusted by the user.

The Goodness-of-Fit Test

The goodness of fit between the observed values of sole and intercropped maize yield for each irrigation treatment and the corresponding predicted values from the SALTMED model was evaluated using the coefficient of determination (R^2), root mean square error (RMSE), and Willmott index of agreement (*d-stat*) as follows:

Coefficient of Determination

The coefficient of determination (R^2) value is derived from the total sum of squares, specifically representing the sum of the squared deviations of the original data from the mean. R^2 assumes values ranging from 0 to 1.0, where a value of 1.0 signifies a perfect correspondence, and a value of 0 denotes complete disagreement. R^2 indicates the number of points that lie on the regression line (Jamieson et al., 1998).

$$R^2 = 1 - \frac{\sum_{i=1}^{n} (O_i - S_i)^2}{\sum_{i=1}^{n} (O_i - \bar{O})^2} \tag{8.11}$$

In this context, O_i denotes the observed values, S_i signifies the predicted values, n indicates the quantity of observed and predicted values utilized for comparison, and ($\bar{O}$) represents the mean of the observed values.

Root Mean Square Error

Root mean square error (RMSE) is the standard deviation of the residuals (prediction errors). Residuals quantify the deviation of data points from the regression line. It is commonly employed to quantify the disparity between values predicted by a model and those empirically seen in the experiment being modeled (Jamieson et al., 1998).

$$\text{RMSE} = \sqrt{\left(\frac{\sum_{i=1}^{n} (S_i - O_i)^2}{n} \right)} \tag{8.12}$$

Willmott Index of Agreement

Willmott index of agreement (*d-stat*) is the standardized metric for the extent of model prediction error, ranging from 0 to 1.0. A number of 1.0 signifies a perfect match, while a value of 0 denotes complete disagreement (Willmott, 1981).

$$d-stat = 1 - \frac{\sum_{i=1}^{n} (O_i - S_i)^2}{\sum_{i=1}^{n} \left[\left(|(S_i - \bar{O})| + |(O_i - \bar{O})| \right)^2 \right]} \tag{8.13}$$

Calibrated and Validated Maize Yield under its Intercropping Systems with Cowpea

The outcomes of the goodness of fit test for both calibrated and validated values of sole and intercropped maize considering various cowpea planting dates and distinct irrigation regimes across both growing seasons are displayed in Table 8.13. The results in the table demonstrated that the SALTMED model effectively simulated the growth environment of sole and intercropped maize in the field experiments with a high degree of accuracy, as evidenced by R^2 values of 0.9878 and 0.9979 during the calibrated and validated growing seasons, respectively, along with *d-stat* values of 0.9706 and 0.9711 for the same periods, and low RMSE values of 0.1406 and 1.433 ton/ha in the calibrated and validated growing seasons, respectively.

A notable positive correlation ($p < 0.01$) was seen between the measured yield of the sole and intercropped maize and the corresponding simulated values generated by the SALTMED model for each analyzed growing season (Fig. 8.4).

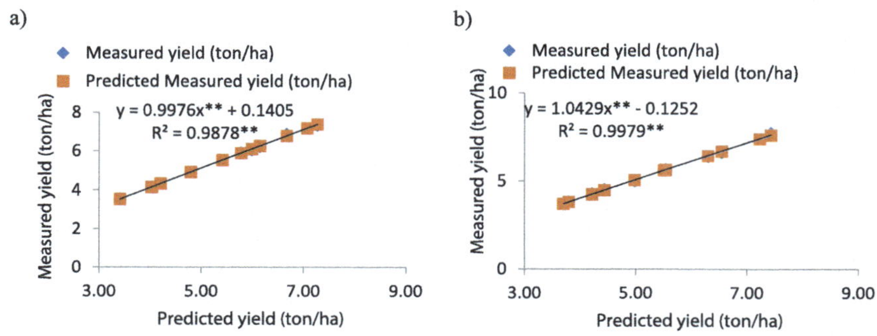

Fig. 8.4 Line fit between the measured yield of sole and intercropped maize and the corresponding simulated values from the SALTMED model in the first growing season (**a**) and in the second growing season (**b**)

Assessment of Climate Change Impact

The CMIP6 mean projections for SSP5–8.5 future climate scenarios were obtained from "The World Bank Group's Climate Change Knowledge Portal (CCKP)" at the following website: https://climateknowledgeportal.worldbank.org/. The CCKP functions as a premier center for global climate data, information, and tools to facilitate decision-making and enhance comprehension of future climates and anticipated dangers. The CCKP was established to facilitate open access to climate data and associated information, enabling users to comprehend future climate scenarios and assess potential risks and impacts using data compiled at national, sub-national, and watershed levels (World Bank Group, 2021).

The climate projection data on that website is modeled data collected from the Coupled Model Intercomparison Projects (CMIP). CMIP data serves as the core dataset for the IPCC Assessment Reports; CMIP6 underpins the IPCC's Sixth Assessment Report, which was initially released in August 2021. The projection data on the website is displayed at a spatial resolution of $1° \times 1°$ (100 km x 100 km). CMIP6 delineates future climatic scenarios via five Shared Socioeconomic Pathways (SSPs): SSP1-1.9, SSP1-2.6, SSP2-4.5, SSP3-7.0, and SSP5-8.5, each representing distinct societal development trajectories. The cumulative radiative forcing levels by 2100, reflecting the total greenhouse gas emissions from all sources, are indicated at the conclusion of each pathway (i.e., −1.9, −2.6, −4.5, 7.0, 8.5) (World Bank Group, 2021). We selected SSP5–8.5 future climate scenario, which expressly envisions a future characterized by heightened exploitation of fossil fuel resources, with increasingly integrated global markets fostering innovations and technical advancements (World Bank Group, 2021). We acquired the predicted monthly data, which includes average maximum and minimum temperatures and relative humidity for Egypt and its sub-region in 2080–2099, corresponding to SSP5–8.5 future climate scenarios derived from a multi-model ensemble. We extracted the monthly data for Minia governorate, where the experiment under study was conducted.

The downloaded data comprises solely monthly records of maximum and minimum temperatures and relative humidity. The SALTMED model employs daily data; hence, the monthly data was transformed into daily data with the MS-Excel function "Random number generation." The model necessitates the input of relative humidity, solar radiation, sunshine duration, and wind speed to compute evapotranspiration and dry matter buildup. The model derived solar radiation estimates from sunshine hours, which were previously determined from maximum and minimum temperatures. Regarding wind speed, Allen et al. (1998) asserted that "the monthly average variation in wind speed is relatively minor and oscillates around mean values, and average wind speed estimates may be derived from regional climate data." Consequently, the current wind speed data was utilized to replace its absence in future instances.

Calculation of Growing Degree Days

To assess the impact of climate change on the future season duration of maize, we utilized the "Degree days, fixed sowing date" tool in SALTMED, which requires the input of stage-specific growth degree days (GDD). This function enables the model to ascertain the commencement and conclusion of each growth phase. The growing degree days (GDD) for each maize growth stage were computed using the equation provided by Yang et al. (1995):

$$\text{GDD} = T_{\text{mean}} \,/\, T_{\text{base}} \tag{8.14}$$

where: T_{base} is base temperature, the critical threshold below which plant growth halts.

Comparison between Present and Projected Average Temperature

The monthly mean temperatures for 2022 and 2023 were juxtaposed with the expected monthly temperatures derived from SSP5-8.5 future climate scenario for the period of 2080–2099, as seen in Fig. 8.5. The figure indicates that throughout the 2080–2099 period, mean temperature values are anticipated to rise, with the projected mean temperature under SSP1-8.5 highly exceeding the recorded values of 2022 and 2023. These results indicated that heat stress will prevail and will adversely affect maize productivity, whether cultivated alone or in intercropping systems.

Projection of Sole and Intercropped Maize Yield under Future Climate

The anticipated maize yields, whether sole cultivated or intercropped under SPP5–8.5 throughout the period of 2080–2099, are expected to be highly reduced (Tables 8.14 and 8.15). The table indicates that under irrigation with 120% ETc, the

Fig. 8.5 Documented average temperature in 2022 and 2023, together with anticipated average temperature under SSP5-8.5 future scenario for the periods 2080–2099 time interval

Table 8.14 Measured versus predicted maize yield (ton/ha) under sole and intercropped systems and irrigation treatments in both growing seasons, and percentage of yield reduction (PR%)

		First growing season (Calibrated)			Second growing season (Validated)		
		Measured	Predicted	PR (%)	Measured	Predicted	PR (%)
RI	Sole maize	6.91	6.67	3	6.64	6.55	1
	PD1	5.56	5.42	3	5.68	5.54	2
	PD2	7.21	7.07	2	7.42	7.23	3
	PD3	7.34	7.27	1	7.76	7.43	4
WS1	Sole maize	5.93	5.78	3	5.63	5.51	2
	PD1	4.92	4.80	2	5.01	4.98	1
	PD2	6.06	5.99	1	6.39	6.3	1
	PD3	6.23	6.15	1	6.62	6.55	1
WS2	Sole maize	4.15	4.03	3	3.85	3.80	1
	PD1	3.54	3.41	4	3.72	3.70	1
	PD2	4.18	4.04	3	4.32	4.22	2
	PD3	4.35	4.21	3	4.55	4.44	2
R^2		0.9878			0.9979		
d-stat		0.9706			0.9711		
RMSE (ton/ha)		0.1406			0.1433		

RI, WS_1, and WS_2 = irrigation with 120, 100, and 80% ETc, respectively; PD1 = cowpea sown 2 weeks prior to maize cultivation; PD2 = cowpea sown concurrently with maize cultivation; PD3 = cowpea sown 2 weeks subsequent to maize cultivation

Table 8.15 Measured versus projected maize sole and intercropped yield (ton/ha) in the two growing seasons, and percentage of yield reduction (PR%) under SSP5-8.5 future scenario in 2080–2099

		First growing season			Second growing season		
		Measured	Predicted	PR (%)	Measured	Predicted	PR (%)
RI	Sole maize	6.91	5.41	22	6.64	5.08	23
	PD1	5.56	4.51	19	5.68	4.85	15
	PD2	7.21	6.09	16	7.42	6.42	13
	PD3	7.34	6.32	14	7.76	6.93	11
WS1	Sole maize	5.93	4.35	27	5.63	4.01	29
	PD1	4.92	3.76	24	5.01	3.86	23
	PD2	6.06	4.78	21	6.39	4.99	22
	PD3	6.23	4.99	20	6.62	5.25	21
WS2	Sole maize	4.15	2.49	40	3.85	2.25	42
	PD1	3.54	2.22	37	3.72	2.41	35
	PD2	4.18	2.67	36	4.32	2.83	34
	PD3	4.35	2.89	34	4.55	3.04	33
R^2		0.9816			0.9725		
d-stat		0.7974			0.8117		
RMSE (ton/ha)		1.4005			1.4288		

RI, WS_1, and WS_2 = irrigation with 120, 100, and 80% ETc, respectively; PD1 = cowpea planted 2 weeks before maize cultivation; PD2 = cowpea planted in the same date as maize cultivation; PD3 = cowpea planted 2 weeks after maize cultivation

most significant yield loss in sole maize is due to the detrimental effects of continuous maize cultivation on soil health, including nutrient depletion, diminished organic matter, and compromised soil structure (Fujisao et al., 2020), with exacerbated losses anticipated in the subsequent growing season. Moreover, the anticipated soil degradation due to climate change in the future, along with the increase in air temperature, would exacerbate yield harm.

A lesser decrease in the yield of intercropped maize compared to that of sole maize will occur. Moreover, these losses will be diminished in the second growing season relative to the first, owing to the residual effect of cowpea during the second growing season. Cowpea cultivation is believed to confer substantial residual benefits to subsequent crops by enriching the soil with nitrogen and enhancing soil quality and structure (Abdul Rahman et al., 2022).

A greater decrease in sole maize yield is anticipated under water stress conditions, with the most significant reduction occurring under irrigation with 80% ETc. The decline in the production of intercropped maize will be less than that of sole maize, with reduced losses in the second growing season relative to the first. The decline in the yield of intercropped maize will be most pronounced under irrigation with 80% ETc, where the maize plants will experience both heat stress resulted from temperature rise, along with the imposed water stress. This cumulative impact of heat stress along with water stress will result in yield reductions for sole maize of 40% and 42% during the first and second growing seasons, respectively, under irrigation with 80% ETc. Lower yield losses in intercropped maize were observed, particularly when cowpea was intercropped 2 weeks post maize cultivation, even under severe water constraint, namely irrigation with 80% ETc. The intercropped maize yield experience losses of 34% and 33% in the first and second seasons, respectively, when cowpea was intercropped 2 weeks following maize cultivation under irrigation with 80% ETc (Table 8.15).

The goodness of fit between the measured yield of sole and intercropped maize and the predicted yield exhibited greater deviation, with R^2 values of 0.9816 and 0.9725 in the first and second seasons, respectively. The *d-stat* values were 0.7974 and 0.8117 in the first and second seasons, respectively, while the RMSE values were 1.4005 and 1.4288 ton/ha (Table 8.15).

Numerous research have projected future maize productivity in Egypt under climate change utilizing simulation models like CropSyst and SALTMED. Ouda et al. (2009) indicated that maize cultivated in clay soil with surface irrigation is projected to experience significant yield declines approximately 49% under the A2 scenario and 47% under the B2 scenario for the year 2040. Conversely, Ouda et al. (2012) determined that maize cultivated on silty clay soil with drip irrigation will incur reduced yield losses, estimated at 28% under A2 scenario and 25% under B2 scenario for the year 2030. Morsy (2015) anticipated a 13–15% reduction in maize output for the year 2030, utilizing the RCP6.0 scenario generated by the CCSM4 model. A recent study by Ouda and Zohry (2024) employed the SALTMED model to project the impact of future climate scenarios on the yield of sole maize and intercropped maize with cowpea. They reported a decrease in the yield of sole maize by 12–25% and a reduction of 7–10% for intercropped maize under adequate irrigation conditions.

Conclusion

This study demonstrates that cowpea–maize intercropping provides substantial agronomic, environmental, and economic benefits in the arid conditions of Middle Egypt. Across two growing seasons, the intercropping systems consistently outperformed sole maize in land and water use efficiencies, and overall profitability. Planting cowpea either simultaneously with maize or 2 weeks after maize optimized maize yield minimized interspecific competition, and enhanced total system productivity. Conversely, earlier cowpea planting by 2 weeks maximized cowpea biomass and contributed the greatest increases in soil organic matter and available nitrogen and phosphorus.

Intercropping also generated important regulatory ecosystem services. Increased soil organic matter, enhanced nutrient cycling through biological nitrogen fixation, and greater availability of phosphorus improved soil fertility and long-term sustainability. The system moderated the crop microclimate, reducing soil evaporation, improving moisture retention, and alleviating heat stress effects—attributes that became increasingly valuable under water deficit conditions. The intercrop canopy and early ground cover contributed to reduced weed biomass and further strengthened resource-use efficiency.

Water use analyses showed that intercropping increased total water consumption but substantially improved water use efficiency and water-equivalent ratios, confirming that each unit of water produced more biomass in the intercrop than in monocultures. Even under moderate and severe water stress, cowpea–maize systems sustained higher productivity than sole maize, with water productivity maximized under 100% ETc irrigation.

The SALTMED model proved effective in simulating intercrop performance, with strong calibration and validation statistics. Its application to future climate scenarios revealed that maize yields will be negatively affected by projected heat increases under SSP5-8.5. However, intercropping with cowpea mitigated these losses, particularly when cowpea was planted after maize. The residual improvements to soil fertility and microclimatic buffering helped maintain higher yields under future climate conditions compared with monoculture maize.

Overall, cowpea–maize intercropping emerges as a climate-smart strategy capable of enhancing agricultural resilience, improving soil health, increasing water and nitrogen efficiency, and ensuring higher economic returns. These findings support the adoption and scaling of legume–cereal intercropping as a sustainable production system for arid and semi-arid regions of Egypt and similar environments worldwide.

References

Abd El-Wahed, M. H., & Snyder, R. (2015). Calculating sunshine hours and reference evapotranspiration in arid regions when solar radiation data are limited. *Irrigation and Drainage*. https://doi.org/10.1002/ird.1920

Abdul Rahman, N., Larbi, A., Berdjour, A., et al. (2022). Cowpea living mulch effect on soil quality and grain yield in smallholder maize-based cropping system of northern Ghana. *Journal of Soil Science and Plant Nutrition, 22*, 3925–3940. https://doi.org/10.1007/s42729-022-00942-5

Abou-Keriasha, M. A., Ibrahim, S. I., & Mohamadain, E. E. A. (2011). Effect of cowpea intercropping date in maize and sorghum fields on productivity and infestation weed. *Egyptian Journal of Agronomy, 33*(1), 35–49.

Akchaya, K., Parasuraman, P., Pandian, K., Vijayakumar, S., Thirukumaran, K., Mustaffa, M. R. A. F., Rajpoot, S. K., & Choudhary, A. K. (2025). Boosting resource use efficiency, soil fertility, food security, ecosystem services, and climate resilience with legume intercropping: A review. *Frontiers in Sustainable Food Systems, 9*, 1527256. https://doi.org/10.3389/fsufs.2025.1527256

Allen, R. G., Pereira, L. S., Raes, D., & Smith, M. (1998). *Crop evapotranspiration: Guideline for computing crop water requirements* (pp. 109–158). FAO. No56.

Amorim, M. R., Mendes, L. W., Antunes, J. E. L., Oliveira, L. M. S., Melo, V. M. M., Oliveira, F. A. S., Aquino, J. P. A., Rocha, S. M. B., Pereira, A. P. A., da Costa, A. F., Oliveira, J. P., Araujo, F. F., Figueiredo, M. V. B., & Araujo, A. S. F. (2022). Cowpea nodules host a similar bacterial community regardless of soil properties. *Applied Soil Ecology, 172*. https://doi.org/10.1016/j.apsoil.2021.104354

Arshad, M. (2025). Experimental assessment of carbon sequestration in sole and intercropping of maize (*Zea mays* L.) with legumes in mountain agroecosystem. *Agronomy Journal, 117*, e70131. https://doi.org/10.1002/agj2.70131

Ayele, H. M. (2020). Evaluation of the effect of maize-legume intercropping on soil moisture improvement in arid area of Bena-Tsemay district, South Omo zone, Southern Ethiopia. *International Journal of Agricultural Research, Innovation and Technology, 10*, 80–86.

Balík, J., Kulhánek, M., Černý, J., Sedlář, O., & Suran, P. (2020). Soil organic matter degradation in long-term maize cultivation and insufficient organic fertilization. *Plants, 9*, 1217. https://doi.org/10.3390/plants9091217

Banik, P., Midya, A., Sarkar, B. K., & Ghose, S. S. (2006). Wheat and chickpea intercropping systems in an additive series experiment: Advantages and weed smothering. *European Journal of Agronomy, 24*, 325–332.

Bhatti, I. H., Ahmad, R., Jabbar, A., Nadeem, M., Khan, M. M., & Vains, S. N. (2013). Agronomic performance of mash bean as an intercrop in sesame under different planting patterns. *Emirates Journal of Food and Agriculture, 25*, 52–57.

Boukar, O., Bhattacharjee, R., Fatokun, C., Kumar, P. L., & Gueye, B. (2013). Cowpea. In M. Singh, H. D. Upadhyaya, & I. S. Bisht (Eds.), *Genetic and genomic resources of grain legume improvement* (pp. 137–155).

Collins, H. P., Fay, P. A., Kimura, E., Fransen, S., & Himes, A. (2017). Intercropping with switchgrass improves net greenhouse balance in hybrid poplar plantations on a sand soil. *Soil Science Society of America Journal, 81*, 781–795.

Dahmardeh, M., Ghanbari, A., Syahsar, B. A., & Ramrodi, M. (2010). The role of intercropping maize (Zea mays L.) and cowpea (Vigna unguiculata L.) on yield and soil chemical properties. *African Journal of Agricultural Research., 5*(8), 631–636.

Dhakal, Y., Meena, R. S., & Kumar, S. (2016). Effect of INM on nodulation, yield, quality and available nutrient status in soil after harvest of green gram. *Legume Research, 39*(4), 590–594.

Dhima, K. V., Lithourgidis, A. A., Vasilakoglou, I. B., & Dordas, C. A. (2007). Competition indices of common vetch and cereal intercrops in two seeding ratios. *Field Crops Research, 100*, 249–256.

Dimande, P., Arrobas, M., & Rodrigues, M. Â. (2024). Intercropped maize and cowpea increased the land equivalent ratio and enhanced crop access to more nitrogen and phosphorus compared to cultivation as sole crops. *Sustainability, 16*, 1440. https://doi.org/10.3390/su16041440

Drăghici, R., Nanu, Ş., Diaconu, A., Bîrsoghe, C., Netcu, F., Paraschiv, A. N., & Dima, M. (2024). Effect of green cowpea manure on the quality properties of sandy soils under the specific conditions of the dry climate in Romania. *Agronomy, 14*(11), 2474. https://doi.org/10.3390/agronomy14112474

Dube, E. D. N., Madanzi, T., Kapenzi, A., & Masvaya, E. (2014). Root length density in maize/cowpea intercropping under a basin tillage system in a semi-arid area of Zimbabwe. *American Journal of Plant Sciences, 5*, 1499–1507. https://doi.org/10.4236/ajps.2014.511165

Dwivedi, A., Dev, I., Kumar, V., Yadav, R. S., Yadav, M., & Gupta, D. (2015). Potential role of maize-legume intercropping systems to improve soil fertility status under smallholder farming systems for sustainable agriculture in India. *International Journal of Life Sciences Biotechnology and Pharma Research, 4*, 145–157.

El-Ghobashy, Y., Shams, A., & Lamlom, M. (2018). Maximizing land use efficiency by intercropping cowpea with some maize cultivars under different maize planting geometries. *Agricultural Sciences, 9*, 1601–1620.

El-Ghobashy, Y., Elmehy, A., & El-Douby, K. (2020). Influence of intercropping cowpea with some maize hybrids and N nano-mineral fertilization on productivity in salinity soil. *Egyptian Journal of Agronomy, 42*(1), 63–78. https://doi.org/10.21608/agro.2020.19752.1194

Elsaid, S. M., Elmorshedy, M. A., Galal, A. H., Abdel Motagally, F. M. F., & Abdullah, M. A. M. (2019). Effect of intercropping maize with cowpea on the yield and its quality. *Assiut Journal of Agricultural Sciences, 50*(3), 39–47.

El-Shamy, M. A., El-Mansoury, M. A. M., El Bialy, M. A., Helmy, M. H., & Hassan, A. F. (2021). Response of maize to different intercropping systems and different sowing dates of cowpea. *Alinteri Journal of Agricultural Sciences, 36*(1), 169–180. https://doi.org/10.47059/alinteri/V36I1/AJAS21025

El-Shamy, M. A., EL-Gaafarey, T. G., Sayed, M. R. I., & ElSanosy, H. M. (2022). Evaluating the efficiency of intercropping systems for three maize hybrids with cowpea on yield productivity and net profit. *Middle East Journal of Agriculture Research, 11*(04), 1050–1063.

Eskandari, H. (2012). Intercropping of maize (Zea mays L.) with cowpea (Vigna sinensis) and mung bean (Vigna radiate): Effects of complementarity of intercrop components on resources consumption, dry matter production and legume forage quality. *Journal of Basic and Applied Science Research, 2*, 355–360.

Eskandari, H., & Kazemi, K. (2011). Weed control in maize-cowpea intercropping system related to environmental resources consumption. *Not Sci Biol, 3*(1), 57–60.

FAOSTAT. (2025). https://www.fao.org/faostat/en/#data

Ferguson, B. J., Lin, M. H., & Gresshoff, P. M. (2013). Regulation of legume nodulation by acidic growth conditions. *Plant Signaling & Behavior, 8*, e23426.

Fujisao, K., Khanthavong, P., Oudthachit, S., et al. (2020). Impacts of the continuous maize cultivation on soil properties in Sainyabuli province, Laos. *Scientific Reports, 10*(11231). https://doi.org/10.1038/s41598-020-67830-9

Gaspar, J. D. C., Furtado, M. B., Silva, W. J. D. S., Reis, I. D. S., Machado, N. A. F., de Farias, F., Furtado, J. L. B., de Andrade, H. A. F., Sobral, R. D. S., Parra-Serrano, L. J., Rodrigues, K. D. M., & Silva-Matos, R. R. S. (2018). Impacts of residual phosphorus on the production of cowpea in the Cerrado Region. *American Journal of Plant Sciences, 9*, 645–658. https://doi.org/10.4236/ajps.2018.94051

Gebru, H. A. (2015). Review on the comparative advantage of intercropping systems. *Journal of Biology, Agriculture and Healthcare., 6*, 2.

Ghanbari, A., Dahmardeh, M., Siahsar, B. A., & Ramroudi, M. (2010). Effect of maize (Zea mays L.) - cowpea (Vigna unguiculata L.) intercropping on light distribution, soil temperature and soil moisture in arid environment. *Journal of Food, Agriculture & Environment, 8*(1), 102–108.

Ghosh, P. K. (2004). Growth, yield, competition and economics of groundnut/cereal fodder intercropping systems in the semi-arid tropics of India. *Field Crops Research, 88*, 227–237.

Ghosh, P. K., Bandyopadhyay, K. K., Wanjari, R. H., Manna, M. C., Misra, A. K., Mohanty, M., et al. (2007). Legume effect for enhancing productivity and nutrient use-efficiency in major cropping systems–an Indian perspective: A review. *Journal of Sustainable Agriculture, 30*, 59–86. https://doi.org/10.1300/J064v30n01_07

Gomaa, M. A., Kandil, E. E., El-Dein, A. A. M. Z., et al. (2021). Increase maize productivity and water use efficiency through application of potassium silicate under water stress. *Scientific Reports, 11*, 224. https://doi.org/10.1038/s41598-020-80656-9

Gomez, K. A., & Gomez, A. A. (1984). *Statistical procedures for agricultural research* (2nd ed., pp. 22–67). John Wiley and Sons.

Gustiningsih, D., Purnamawati, H., Lubis, I., Syukur, M., Kondo, T., & Higuchi, H. (2023). Response of cowpea growth, yield and organic acid secretion in acidic soil to variability in population and minus one element fertilizer test. *Journal of Tropical Crop Science, 10*(03), 196–204. https://doi.org/10.29244/jtcs.10.03

Hamd-Alla, W. A., Shalaby, E. M., Dawood, R. A., & Zohry, A. A. (2014). Effect of cowpea (Vigan sinensis L.) with maize (Zea mays L.) intercropping on yield and its components. *International Scholarly and Scientific Research & Innovation, 8*(11), 1170–1176.

Haruna, A., Kombiok, J. M., Mohamed, A. M., Sarkodie-Addo, J., Larbi, A., & Rahman, N. A. (2018). Profitability of cowpea intercropped with maize in West Africa Guinea savanna. *The Journal of Agricultural Science, 10*, 185. https://doi.org/10.5539/jas.v10n11p185

He, H., Yang, L., Fan, L., Zhao, L., Wu, H., et al. (2011). The effect of intercropping of maize and soybean on microclimate. In *5th computer and computing Technologies in Agriculture (CCTA), Beijing, China* (pp. 257–263) ff10.1007/978-3-642-27278-3_27ff.ffhal-01360987f.

He, H., Yang, L., Fan, L., Zhao, L., Wu, H., Yang, J., & Li, C. (2012). The effect of inter-cropping of maize and soybean on microclimate. In Li & Chen (Eds.), *Computer and Computing Technologies in Agriculture V. CCTA 2011. IFIP advances in information and communication technology* (pp. 257–263). Springer.

Hesse, P. R. A. (1971). *Text book of soil chemical analysis* (pp. 27–45). Juan Murry (Publisher) Ltd..

Hu, H.-Y., Li, H., Hao, M.-M., Ren, Y. N., Zhang, M. K., Liu, R. Y., Zhang, Y., Li, G., Chen, J. S., Ning, T. Y., et al. (2021). Nitrogen fixation and crop productivity enhancements co-driven by intercrop root exudates and key rhizosphere bacteria. *Journal of Applied Ecology, 58*(10), 2243–2255. https://doi.org/10.1111/1365-2664.13964

Israelsen, O. W., & Hansen, V. E. (1962). *Irrigation principles and practices* (3rd ed.). Wiley International Edition.

Jackson, M. L. (1973). *Soil chemical analysis*. Prentice-Hall, Inc..

Jamieson, P. D., Porter, J. R., Goudriaan, J., Ritchie, J. T., Keulen, H., & Stol, W. (1998). A comparison of the models AFRCWHEAT2, CERES-Wheat, Sirius, SUCROS2 and SWHEAT with measurements from wheat grown under drought. *Field Crops Research, 55*, 23–44.

Jamshidi, K., Yousefi, A. Z., & Oveisi, M. (2013). Effect of cowpea (Vigna unguiculata) intercropping on weed biomass and maize (Zea mays) yield. *New Zealand Journal of Crop and Horticultural Science, 41*(4), 180–188. https://doi.org/10.1080/01140671.2013.807853

Jangir, C. K., Kumar, S., Lakhran, H., & Meena, R. S. (2017). Towards mitigating malnutrition in pulses through biofortification. *Trends in Biosciences, 10*(17), 2999–3002.

Jensen, E. S., Ambus, P., Bellostas, N., Boisen, S., Brisson, N., Corre Hellou, G., Crozat, Y., Dahlmann, C., Dibet, A., von Fragstein, P., & Gooding, M. (2006). Intercropping of cereals and grain legumes for increased production, weed control, improved product quality and prevention of N–losses in European organic farming systems, *International conferences: Joint organic congress—theme 4: Crop systems and soils*.

Katsaruware, R. D., & Manyanhaire, I. O. (2009). Maize-cowpea intercropping and weed suppression in leaf stripped and detasselled maize in Zimbabwe. *Electronic Journal of Environmental, Agricultural and Food Chemistry. 8*(11), 1218–1226.

Klute, A. (1986). *Methods of soil analysis, Part I. physical and mineralogical method*. American Society of Agronomy.

Kumar, N., Hazra, K. K., Nath, C. P., Praharaj, C. S., & Singh, U. (2018). Grain legumes for resource conservation and agricultural sustainability in South Asia. In *Legumes for soil health and sustainable management* (pp. 77–107). Springer.

Kumar, S., Meena, R. S., Datta, R., Verma, S. K., Yadav, G. S., Pradhan, G., Molaei, A., Rahman, G. M., & Mashuk, H. A. (2020). Legumes for car bon and nitrogen cycling: An organic approach. In *Carbon and Nitrogen Cycling in Soil* (pp. 337–375). Springer.

Kumawata, A., Bamboriyab, S. D., Meenac, R. S., Yadavd, D., et al. (2022). Legume-based intercropping to achieve the crop, soil, and environmental health security. In *Advances in legumes for sustainable intensification* (pp. 307–328). Academic Press. https://doi.org/10.1016/B978-0-323-85797-0.00005-7

Lal, R. (2011). Sequestering carbon in soils of agro-ecosystems. *Food Policy, 36*, S33–S39.
Lal, R. (2017). Improving soil health and human protein nutrition by pulses based cropping systems. *Advances in Agronomy, 145*, 167–204.
Latati, M., Blavet, D., Alkama, N., Laoufi, H., Drevon, J.-J., et al. (2014). The intercropping cowpea-maize improves soil phosphorus availability and maize yields in an alkaline soil. *Plant and Soil, 385*(1-2), 181–191. https://doi.org/10.1007/s11104-014-2214-6. ⟨hal-01268708⟩.
Layek, J., Das, A., Mitran, T., Nath, C., Meena, R. S., Yadav, G. S., et al. (2018). Cereal+ legume intercropping: An option for improving productivity and sustaining soil health. In *Legumes for soil health and sustainable management* (pp. 347–386).
Liu, H., Gao, X., Li, C., Cai, Y., Song, X., & Zhao, X. (2025). Intercropping increases plant water availability and water use efficiency: A synthesis. *Agriculture, Ecosystems & Environment, 379*, 109360. https://doi.org/10.1016/j.agee.2024.109360
Mansour, H. (2025). Enhancing maize productivity through AQUACROP modeling under modern irrigation systems and drought stress conditions. *Egyptian Journal of Agronomy, 47*(1), 55–71.
Mao, L. L., Zhang, L. Z., Li, W. W., Werf, W. V. D., Sun, J. H., Spiertz, H., & Li, L. (2012). Yield advantage and water saving in maize/pea intercrop. *Field Crops Research, 138*, 11–20.
Masoued, S. R. (2017). *Climate and its impact on diseases in the Northern Upper Egypt (applied climatology study)*. Msc Thesis. (pp. 15–36). Department of Geography, Cairo University, Egypt.
Megawer, E. A., Sharaan, A. N., & El-Sherif, A. M. (2010). Effect of intercropping patterns on yield and its components of barley, lupine or chickpea grown in newly reclaimed soil. *Egyptian Journal of Applied Science, 25*(9), 437–452.
Melo, A. A. R., Araújo, M. A., Mendes, N. A. C., & Reis, A. R. (2025). Drought stress disrupts biological nitrogen fixation and starch accumulation compromising growth and yield of cowpea plants. *Plant Physiology and Biochemistry, 224*, 109931. https://doi.org/10.1016/j.plaphy.2025.109931
Mishra, K. (2019). Evaluation of maize cowpea intercropping as fodder through front line demonstration. *Journal of Medicinal Plants Studies, 7*, 82–85.
Mitran, T., Meena, R. S., Lal, R., Layek, J., Kumar, S., Datta, R., et al. (2018). Role of soil phosphorus on legume production. In Meena et al. (Eds.), *Legumes for soil health and sustainable management*. Springer. https://doi.org/10.1007/978-981-13-0253-4_15
Morris, R. A., & Garrity, D. P. (1993). Resource capture and utilization in intercropping: Water. *Field Crops Research, 34*, 303–317.
Morsy, M. (2015). *Use of regional climate and crop simulation models to predict wheat and maize productivity and their adaptation under climate change*. PhD thesis. Faculty of Science Al-Azhar University.
Moursi, E., Aziz, M. A., Aiad, M., & Darwesh, R. (2013). Effect of water stress, biofertilizers and nitrogen application rates on cowpea yield and some water relations in the north middle Nile Delta region. *Journal of Soil Sciences and Agricultural Engineering, 4*(11), 1289–1311. https://doi.org/10.21608/jssae.2013.52899
Mugi-Ngenga, E., Bastiaans, L., Anten, N., Zingore, S., Baijukya, F., & Giller, K. E. (2023). The role of inter-specific competition for water in maize-legume intercropping systems in northern Tanzania. *Agricultural Systems, 207*, 103619.
Nasar, J., Khan, W., Khan, M. Z., Gitari, H. I., Gbolayori, J. F., Moussa, A. A., et al. (2021). Photosynthetic activities and photosynthetic nitrogen use efficiency of maize crop under different planting patterns and nitrogen fertilization. *Journal of Soil Science and Plant Nutrition, 21*, 2274–2284. https://doi.org/10.1007/s42729-021-00520-1
Nemecek, T., von Richthofen, J.-S., Dubois, G., Casta, P., Charles, R., & Pahl, H. (2008). Environmental impacts of introducing grain legumes into European crop rotations. *European Journal of Agronomy, 28*, 380–393.
Ouda, S., & Zohry, A. (2023). Increasing sustainability by intercropping legume crops with sugar beet under imposed water stress. *Irrigation and Drainage*, 1–26. Available from:. https://doi.org/10.1002/ird.2884

Ouda, S., & Zohry, A. (2024). Increasing land and water use efficiencies of wheat: Case study of Egypt. In *Integration of legume crops with cereal crops under changing climate: Sustainably increases food production*. Springer Publishing House. ISBN: 978-3-031-68101-1.

Ouda, S., & Zohry, A. (2026). Optimizing resource utilization through intercropping soybeans (Glycine max (L.) Merr.) and maize (Zea mays L.) in Middle Egypt. *Acta agriculturae Slovenica. Acta Agriculturae Slovenica, 122*(1), 1–23. https://doi.org/10.14720/aas.2026.122.1.24399. Under review.

Ouda, S. A., Khalil, F. A., & Yousef, H. (2009). Using adaptation strategies to increase water use efficiency for maize under climate change conditions. *Proceeding of 13th international conference on water technology. Egypt.*

Ouda, S., Abdrabbo, M., & Noreldin, T. (2012). Effect of changing sowing dates and irrigation scheduling on maize yield grown under climate change conditions. *10th International Conference of Egyptian Soil Science Society (ESSS) and 4th International Conference of Water Requirements & Metrology Dept. 5–7 November. Ameria, Egypt (Cl 07).*

Ouda, S., Zohry, A., & Abdel-Wahab, S. (2020). Sustainable intensive cropping to reduce irrigation-induced erosion: II. Changing cropping sequence under sprinkler irrigation practice. *Moroccan Journal of Agricultural Science., 1*(1), 26–33.

Pankou, C., Lithourgidis, A., & Dordas, C. (2021). Effect of irrigation on intercropping systems of wheat (Triticum aestivum L.) with pea (Pisum sativum L.). *Agronomy, 11*(2), 283. https://doi.org/10.3390/agronomy11020283

Ragab, R. (2002). A holistic generic integrated approach for irrigation, crop and field management: The SALTMED model. *Environmental Modelling & Software, 17*(4), 345–361.

Ragab, R. (2015). Integrated management tool for water, crop, soil and N-fertilizers: The saltmed model. *Irrigation and Drainage, 64*(1), 1–12.

Ragab, R., Malash, N., Gawad, G. A., Arslan, A., & Ghaibeh, A. (2005a). A holistic generic integrated approach for irrigation, crop and field management: 1. The SALTMED model and its calibration using field data from Egypt and Syria'. *Agricultural Water Management, 78*(1), 67–88.

Ragab, R., Malash, N., Gawad, G. A., Arslan, A., & Ghaibeh, A. (2005b). A holistic generic integrated approach for irrigation, crop and field management: 2. The SALTMED model validation using field data of five growing seasons from Egypt and Syria. *Agricultural Water Management, 78*(1), 89–107.

Rao, M. R., & Willey, R. W. (1980). Evaluation of yield stability in intercropping: Studies on sorghum/pigeon pea. *Experimental Agriculture, 16*(2), 105–116.

Raza, M. A., Gul, H., Wang, J., Yasin, H. S., Qin, R., Bin Khalid, M. H., Naeem, M., et al. (2021). Land productivity and water use efficiency of maize-soybean strip intercropping systems in semi-arid areas: A case study in Punjab Province, Pakistan. *Journal of Cleaner Production, 308*, 127282. https://doi.org/10.1016/j.jclepro.2021.127282

Roohi, M., Arif, M. S., Guillaume, T., & Yasmeen, T. (2022). Role of fertilization regime on soil carbon sequestration and crop yield in a maize-cowpea intercropping system on low fertility soils. *Geoderma, 428*, 116152. https://doi.org/10.1016/j.geoderma.2022.116152

Salama, H. S. A., Nawar, A. I., & Khalil, H. E. (2022). Intercropping pattern and N fertilizer schedule affect the performance of additively intercropped maize and forage cowpea in the Mediterranean Region. *Agronomy, 12*, 107. https://doi.org/10.3390/agronomy12010107

Sánchez-Navarro, V., Zornoza, R., Faz, Á., & Fernández, J. A. (2019). Does the use of cowpea in rotation with a vegetable crop improve soil quality and crop yield and quality? A field study in SE Spain. *European Journal of Agronomy, 107*, 10–17. https://doi.org/10.1016/j.eja.2019.03.007

Senghor, Y., Balde, A. B., Manga, A. G. B., Affholder, F., Letourmy, P., Bassene, C., et al. (2023). Intercropping millet with low-density cowpea improves millet productivity for low and medium N input in semi-arid Central Senegal. *Heliyon, 9*, e17680.

Shehata, M. A., & Hamd-Alla, W. (2023). Response of intercropping cowpea with maize to potassium fertilizer and foliar application of boron on the productivity of both crops. *Journal of Plant Production, 4*(14), 405–411.

Shen, Q. R., & Chu, G. X. (2004). Bi-directional nitrogen transfer in an intercropping system of peanut with rice cultivated in aerobic soil. *Biology and Fertility of Soils, 40*, 81–87.

Singh, B. B. (2014). *Cowpea: The food legume of the 21st century*. Crop Science Society of America.

Singh, B. B., Ajeigbe, H. A., Tarawali, S. A., Fernandez-Rivera, S., & Abubakar, M. (2003). Improving the production and utilization of cowpea as food and fodder. *Field Crops Research, 84*, 169–177.

Stagnari, F., Maggio, A., Galieni, A., et al. (2017). Multiple benefits of legumes for agriculture sustainability: An overview. *Chemical and Biological Technologies in Agriculture, 4*, 2. https://doi.org/10.1186/s40538-016-0085-1

Suhi, A., Mia, S., Khanam, S., Hasan Mithu, M., Uddin, M. K., Muktadir, M. A., Ahmed, S., & Jindo, K. (2022). How does maize-cowpea intercropping maximize land use and economic return? A field trial in Bangladesh. *Land, 11*, 581. https://doi.org/10.3390/land11040581

Taha, A., & Kasem, E. (2022). Water management of maize-cowpea intercropping system under surface irrigation. *Moroccan Journal of Agricultural Sciences, 3*(1), 29–38.

Tamta, A., Kumar, R., Ram, H., Meena, R. K., Kumar, U., Yadav, M. R., et al. (2019). Nutritional portfolio of maize and cowpea fodder under various intercropping ratio and balanced nitrogen fertilization. *The Indian Journal of Animal Sciences, 89*, 276–280. https://doi.org/10.56093/ijans.v89i3.88041

Tan, K. H. (1996). *Soil Sampling, Preparation and Analysis* (p. 391). Marcel Dekker, Inc..

Tharanathan, R. N., & Mahadevamma, S. (2003). Grain legumes–a boon to human nutrition. *Trends in Food Science & Technology, 14*, 507–518.

Vasconcelos, M. W., Grusak, M. A., Pinto, E., Gomes, A., Ferreira, H., Balázs, B., et al. (2020). The biology of legumes and their agronomic, economic, and social impact. In *The plant family Fabaceae* (pp. 3–25). Springer. https://doi.org/10.1007/978-981-15-4752-2_1

Vermeiren, L., & Jopling, G. A. (1984). *Localized irrigation*. FAO irrigation and drainage paper no. 36, Rome, Italy.

Walkley, A., & Black, I. A. (1934). An examination of Degtjareff method for determining organic carbon in soils: Effect of variations in digestion conditions and of inorganic soil constituents. *Soil Science, 63*, 251–263.

Waller, R. A., & Duncan, D. B. (1969). *Bays rule for the symmetric multiple comparison problem*. American State Association.

Wang, Z., Wu, P., Zhao, X., Gao, Y., & Chen, X. (2015). Water use and crop coefficient of the wheat–maize strip intercropping system for an arid region in northwestern China. *Agricultural Water Management, 161*, 77–85. https://doi.org/10.1016/j.agwat.2015.07.012

Wang, W., Li, M-Y., Wang, Y., Li, J-M., Zhang, W., et al., (2025). Legume intercropping improves soil organic carbon stability in drylands: A 7-year experimental validation, Agriculture, Ecosystems & Environment, *381*, 109456, https://doi.org/10.1016/j.agee.2024.109456.

Willmott, C. J. (1981). On the validation of models. *Physical Geography, 2*, 184–194.

World Bank group. (2021). *Climate change knowledge portal (CCKP) user manual* (p. 18). World Bank Group.

Yang, S., Logan, J., & Coffey, D. L. (1995). Mathematical formulae for calculating the base temperature for growing degree days. *Agricultural and Forest Meteorology, 74*(1–2), 61–74. https://doi.org/10.1016/0168-1923(94)02185-M

Zhang, Y., Sun, Z., Su, Z., Du, J., Bai, W., Wang, Q., Wang, R., et al. (2022). Root plasticity and interspecific complementarity improve yields and water use efficiency of maize/soybean intercropping in a water-limited condition. *Field Crops Research, 282*, 108523. https://doi.org/10.1016/j.fcr.2022.108523

Zohry, A. A. (2005). Effect of preceding winter crops and intercropping on yield, yield components and associated weeds in maize. *Annals of Agricultural Science, Moshtohor, 43*(1), 139–148.

Zohry, A., & Ouda, S. (2025). Legume crops enhance water use efficiency under intercropping system with wheat. *Moroccan Journal of Agricultural Sciences, 6*(3), 145–160.

Zohry, A. A., Ouda, S., Hamd-Alla, W., & Shalaby, E. (2017). Evaluation of different crop sequences for wheat and maize in sandy soil. *Acta Agriculturae Slovenica, 109*(2), 383–392.

Chapter 9
Conclusions, Future Perspectives, and Recommendations

Abstract Leguminous crops play a central role in enhancing agroecosystem services and supporting sustainable intensification under changing climatic conditions. This concluding chapter synthesizes the major findings presented throughout the book regarding the multifunctional contributions of legumes to regulating and supporting agroecosystem services. The chapter highlights the role of legumes in improving soil fertility, enhancing nutrient cycling, increasing soil organic carbon sequestration, improving soil conservation, regulating microclimatic conditions, mitigating climate change, and strengthening biological control within agricultural systems. Special emphasis is placed on diversified legume-based systems, including intercropping, crop rotations, sequence cropping, and cover cropping, as effective strategies for improving resource-use efficiency, biodiversity, productivity, and climate resilience. Evidence from Egyptian agroecosystems demonstrates that legumes can substantially improve water productivity, soil health, and crop stability under arid and semi-arid conditions characterized by limited resources and increasing climatic stress. The chapter further discusses future perspectives for legume-based agriculture and identifies research priorities, policy needs, and management strategies required to expand the role of legumes in climate-smart agriculture. Overall, the chapter concludes that legumes should be regarded as foundational components of resilient and sustainable agroecosystems capable of supporting food security, environmental sustainability, and agricultural adaptation under a changing climate.

Keywords Agroecosystem services · Legumes · Sustainable intensification · Climate change · Intercropping · Crop rotation · Soil fertility · Carbon sequestration · Climate-smart agriculture

Introduction

Agricultural systems worldwide are facing unprecedented challenges associated with climate change, land degradation, declining soil fertility, biodiversity loss, and increasing pressure on natural resources. Rising temperatures, irregular rainfall

A. E. Zohry, S. A. H. Ouda, *The Agroecosystem Services of Leguminous Crops*,
https://doi.org/10.1007/978-3-032-32461-0_9

patterns, water scarcity, and more frequent extreme climatic events threaten agricultural productivity and food security, particularly in arid and semi-arid regions. At the same time, the intensive use of synthetic fertilizers, pesticides, and irrigation has contributed to environmental degradation and reduced the long-term sustainability of agricultural systems. Consequently, there is increasing recognition of the need to develop agricultural systems capable of maintaining productivity while preserving ecosystem functioning and enhancing resilience to climatic and environmental stresses.

Within this context, leguminous crops have emerged as key biological components of sustainable and climate-smart agriculture because of their multifunctional contributions to agroecosystem services. Throughout this book, evidence has been presented demonstrating that legumes provide a wide range of regulating and supporting services that improve soil health, enhance nutrient cycling, strengthen biodiversity, improve water-use efficiency, mitigate greenhouse gas emissions, and stabilize agricultural productivity. The incorporation of legumes into cropping systems through intercropping, crop rotations, sequence cropping, and cover cropping enhances agroecosystem functioning while reducing dependence on external inputs.

The findings synthesized across the chapters of this book further demonstrate that legume-based systems are particularly valuable under the environmental conditions prevailing in Egypt and other arid and semi-arid regions, where limited water resources, declining soil organic matter, and increasing climatic variability represent major constraints to agricultural sustainability. Egyptian case studies presented throughout the book clearly illustrate the important role of legumes in improving soil fertility, increasing water productivity, enhancing resilience to heat and drought stress, and supporting sustainable intensification in both old and newly reclaimed agricultural lands.

This concluding chapter integrates the major findings and concepts discussed throughout the book and highlights the critical role of legumes in promoting sustainable agroecosystems under changing climatic conditions. Thus, the objective of this chapter is to synthesize the major conclusions presented throughout the book regarding the role of legumes in enhancing agroecosystem services and supporting sustainable intensification under climate change. The chapter further aims to identify future perspectives, research priorities, policy directions, and management recommendations required to strengthen the integration of legumes into resilient and climate-smart agricultural systems.

Integrated Conclusions

The evidence presented throughout this book demonstrates that leguminous crops constitute one of the most effective biological tools for enhancing agroecosystem services and achieving sustainable intensification under changing climatic conditions. Across diverse agricultural systems, legumes provide multiple regulating and supporting services that improve ecosystem functioning, maintain productivity, and enhance resilience to environmental stresses. Their integration into cropping systems through intercropping, crop rotation, sequence cropping, and cover cropping represents a practical pathway toward reconciling agricultural productivity with environmental sustainability.

A central conclusion emerging from this book is that legumes contribute substantially to the restoration and maintenance of soil fertility through biological nitrogen fixation. The symbiotic relationship between legumes and rhizobia enables the conversion of atmospheric nitrogen into plant-available forms, thereby reducing dependence on synthetic nitrogen fertilizers and improving nutrient-use efficiency. This biological process not only supports the nitrogen requirements of the legume crop itself but also enhances nitrogen availability for companion and subsequent crops. Consequently, legume-based systems contribute to reduced production costs, lower greenhouse gas emissions associated with nitrogen fertilizer manufacture and application, and improve sustainability of agricultural systems.

The book further confirms that legume integration significantly improves soil health and soil quality. Through the production of nitrogen-rich residues and extensive root systems, legumes contribute to increased soil organic matter accumulation, enhanced soil aggregation, improved soil structure, and greater water-holding capacity. These improvements strengthen the resilience of agricultural soils against degradation, erosion, salinization, and water stress. The decomposition of legume residues accelerates nutrient cycling and stimulates soil microbial activity, thereby enhancing long-term soil productivity and ecosystem stability.

Another major finding synthesized throughout the book is the important role of legumes in carbon sequestration and climate change mitigation. Legume-based systems contribute to increased carbon inputs into soil through aboveground biomass, root turnover, and rhizodeposition. The resulting increase in soil organic carbon enhances soil fertility while simultaneously contributing to atmospheric carbon dioxide mitigation. Moreover, the reduced reliance on synthetic nitrogen fertilizers lowers nitrous oxide emissions, further strengthening the climate mitigation potential of legume-based agriculture.

The ecological benefits of legumes extend beyond soil processes to include substantial improvements in biodiversity and biological regulation within agroecosystems. Intercropping and diversified cropping systems involving legumes create more heterogeneous habitats that support beneficial arthropods, pollinators, and soil microorganisms. Such systems contribute to natural pest suppression, improved pollination services, and weed regulation while reducing dependence on synthetic pesticides and herbicides. These ecological interactions strengthen agroecosystem resilience and stability, particularly under conditions of climatic and environmental stress.

The book also highlights the important contribution of legumes to water conservation and climate adaptation. Legume-based systems improve water-use efficiency through enhanced soil structure, increased infiltration, reduced evaporation losses, and improved moisture retention. In intercropping systems, complementary resource use among component crops improves the utilization of water, nutrients, and light while stabilizing yields under drought and heat stress conditions. These benefits are especially important under projected climate change scenarios characterized by increasing temperatures, irregular rainfall, and more frequent extreme weather events.

A major theme consistently emphasized throughout the book is the importance of diversified cropping systems involving legumes for sustainable intensification. Intercropping, crop sequencing, and legume cover cropping improve land-use efficiency, system productivity, and resource-use efficiency while simultaneously enhancing agroecosystem services. Such systems reduce ecological vulnerability associated with monoculture agriculture and contribute to more resilient and environmentally sustainable food production systems.

The Egyptian case studies presented throughout the book provide strong evidence that legume-based cropping systems can effectively address several challenges facing agriculture in arid and semi-arid environments. Under Egyptian conditions, legumes have demonstrated substantial capacity to improve soil fertility, increase soil organic matter, enhance water productivity, stabilize yields, and improve system resilience under heat and water stress conditions. Intercropping systems such as cowpea–maize, soybean–maize, and faba bean–wheat, as well as legume-based rotations and cover cropping strategies, have shown considerable potential for improving productivity while reducing dependence on external inputs.

The evidence synthesized in this book further indicates that the agroecosystem services provided by legumes are strongly influenced by management practices, environmental conditions, crop combinations, planting density, planting date, irrigation management, and residue handling. Consequently, maximizing the benefits of legumes requires site-specific and climate-adaptive management strategies tailored to local agroecological conditions.

Despite the substantial advantages of legumes, the book also identifies important constraints limiting their wider adoption. These include limited availability of high-quality seeds and rhizobial inoculants, insufficient extension services, low farmer awareness regarding agroecosystem services, market instability, and climate-related stresses such as drought, salinity, heat stress, and emerging pests and diseases. In many developing regions, including Egypt, policy support for diversified legume-based systems remains insufficient despite their clear ecological and economic benefits.

Overall, the findings presented throughout this book clearly demonstrate that legumes should not be viewed merely as supplementary crops, but rather as foundational components of sustainable agroecosystems. Their multifunctional contributions to soil fertility, nutrient cycling, climate regulation, biodiversity enhancement, biological control, water conservation, and sustainable productivity position legumes at the center of climate-smart agriculture and sustainable intensification strategies.

Future Perspectives

Future agricultural systems will increasingly require management approaches capable of sustaining productivity while preserving ecosystem integrity under accelerating climate change. In this context, legumes are expected to play an increasingly important role in the transformation of agricultural systems toward greater sustainability and resilience.

Climate change projections indicate that many agricultural regions will experience higher temperatures, increased evapotranspiration, greater rainfall variability, and more frequent drought events. Under these conditions, legume-based systems offer considerable potential for improving climate resilience through enhanced soil moisture conservation, reduced fertilizer dependency, and greater ecosystem stability. Expanding the integration of legumes into agricultural landscapes may therefore become an essential component of adaptation strategies in both irrigated and rainfed systems.

Future agricultural development should move beyond productivity-centered approaches toward multifunctional agroecosystems that simultaneously deliver food production, climate regulation, soil conservation, biodiversity enhancement, and water-use efficiency. Legumes are uniquely positioned to support this transition because of their capacity to provide multiple agroecosystem services simultaneously.

Advances in plant breeding, molecular genetics, biotechnology, and microbial sciences are expected to improve the performance of legumes under stressful environmental conditions. The development of drought-tolerant, heat-tolerant, salinity-tolerant, and pest-resistant legume cultivars will become increasingly important for maintaining productivity under future climate scenarios. Greater understanding of rhizosphere interactions and soil microbial dynamics may also improve the efficiency of biological nitrogen fixation and nutrient cycling.

Digital agriculture, precision farming, and remote sensing technologies may further enhance the management of legume-based systems by improving irrigation scheduling, nutrient management, and crop monitoring. Such technologies can contribute to more efficient resource use while minimizing environmental impacts.

In Egypt and other arid and semi-arid regions, future agricultural sustainability will depend heavily on improving water productivity and maintaining soil health under conditions of limited natural resources. Legume-based systems offer considerable opportunities for improving resource-use efficiency while sustaining agricultural productivity in these vulnerable environments.

Recommendations for Research

- Future research should prioritize long-term multidisciplinary studies evaluating the cumulative effects of legumes on agroecosystem functioning, soil carbon sequestration, nutrient cycling, biodiversity, and climate resilience under diverse environmental conditions.

- Greater research attention should be directed toward optimizing legume-based intercropping systems under varying climatic conditions. Studies are needed to identify the most effective crop combinations, planting densities, spatial arrangements, and planting dates for maximizing productivity and agroecosystem services while minimizing competition between component crops.
- Additional research is required to improve understanding of belowground ecological interactions involving legumes, particularly rhizosphere processes, microbial diversity, nutrient transfer mechanisms, and root system dynamics. Such knowledge is essential for enhancing nutrient-use efficiency and soil biological functioning.
- Research should also focus on improving the efficiency of biological nitrogen fixation through the development of highly effective and locally adapted rhizobial inoculants. Special emphasis should be placed on identifying strains capable of maintaining high nitrogen fixation efficiency under drought, salinity, and high-temperature conditions.
- Climate-resilient legume cultivars should be developed through breeding and biotechnological approaches. Future breeding programs should target enhanced drought tolerance, heat tolerance, salinity tolerance, pest resistance, and greater biomass production while maintaining high nitrogen fixation capacity.
- Further investigations are needed to quantify the contribution of legume-based systems to greenhouse gas mitigation and carbon sequestration under different management practices and agroecological conditions, particularly in arid and semi-arid regions.
- Research on legume-based systems in Egypt should continue expanding, particularly regarding water productivity, conservation agriculture, climate adaptation, and sustainable management of newly reclaimed soils. Greater integration between agronomic, environmental, and socioeconomic research is also required to support wider adoption of legume-based technologies.

Recommendations for Agricultural Practice and Policy

- Agricultural extension services should strengthen farmer awareness regarding the multifunctional benefits of legumes and provide practical guidance on intercropping, crop rotation, inoculant application, residue management, and sustainable soil management practices.
- Governments and agricultural institutions should encourage the wider adoption of legume-based systems through supportive agricultural policies, including subsidies for legume seeds and rhizobial inoculants, financial incentives for diversified cropping systems, and support for sustainable soil management programs.
- The production and distribution of high-quality legume seeds and effective rhizobial inoculants should be improved to facilitate greater adoption by farmers, particularly in developing countries.

- Legume integration should become a central component of climate-smart agriculture initiatives and national climate adaptation and mitigation strategies. Policymakers should recognize the significant contribution of legumes in reducing greenhouse gas emissions, improving soil carbon sequestration, and enhancing resilience to climate variability.
- Conservation agriculture practices involving legumes, residue retention, and legume cover cropping should be promoted to improve soil health, reduce erosion, and enhance water conservation.
- In Egypt, legume-based cropping systems should be incorporated more extensively into strategies addressing soil degradation, declining soil organic matter, water scarcity, and climate stress. Particular attention should be given to promoting cereal–legume intercropping systems, improved fallow systems involving fahl clover, and diversified crop rotations in both old and newly reclaimed lands.
- Strengthening collaboration among researchers, extension specialists, policymakers, and farmers is essential for improving technology transfer and facilitating the practical implementation of legume-based sustainable intensification strategies.

Final Outlook

The future sustainability of agriculture depends on the successful integration of ecological principles into agricultural production systems. Leguminous crops provide a unique opportunity to achieve this objective because of their multifunctional contributions to agroecosystem services and sustainable intensification.

The evidence synthesized throughout this book clearly demonstrates that legumes can improve soil fertility, enhance nutrient cycling, strengthen biodiversity, increase water productivity, mitigate climate change, and stabilize agricultural production under stressful environmental conditions. These functions are especially critical in regions vulnerable to climate variability and natural resource limitations, including Egypt and other arid and semi-arid environments.

The transition toward resilient and sustainable agroecosystems will require greater diversification of cropping systems, reduced dependency on external inputs, stronger emphasis on ecological processes, and broader adoption of climate-smart agricultural practices. Legume-based systems represent one of the most practical and scientifically supported pathways toward achieving these goals.

Ultimately, the strategic integration of legumes into agricultural systems should be considered not merely an agronomic option, but an essential component of future sustainable agriculture. Expanding their role within cropping systems can contribute significantly to global food security, environmental sustainability, and climate resilience while supporting the long-term productivity and stability of agroecosystems under a changing climate.

Final Outlook

The manufacturer's authorised representative in the EU is Springer Nature Customer Service Centre GmbH, Europaplatz 3, 69115 Heidelberg, Germany. If you have any concerns regarding our products, please contact ProductSafety@springernature.com

Printed and bound by CPI Group (UK) Ltd, Croydon, CR0 4YY
21/09/2026
02220008-0002